Proactive Quality Management

cGMP UNIVERSITY
COMPLIANCE EXPERTS

cGMP UNIVERSITY INC.

We lead others follow

Principles of change control

A HANDBOOK FOR QUALITY AND REGULATORY PERSONNEL.

CHANGE CONTROL FROM A QUALITY AND REGULATORY PERSPECTIVE

David N. Muchemu

PROLOGUE

—Making uncontrolled changes is a **violation** of several sections of the **QSR**, including sections 820.30, 820.40, 820.70, 820.75, and 820.181 (From Change Control Procedure section, QSR Manual).

—There is no easy way to properly control changes to devices, processes, device master records, etc. **Change control is a complex process**. Failure to have an adequate change control system can cause equally complex results (from Introduction of QSR Manual).

—These written procedures, including ***any changes*** shall be drafted, reviewed, and approved by the appropriate organizational units (CCB) and reviewed and approved by the Quality control unit (from Subpart F—21 CFR 211.00: Production and Process Controls

—Including *any change* in such specifications, standards, sampling plans, test procedures, or other laboratory control mechanisms shall be drafted by the appropriate organizational unit, and reviewed and approved by the Quality control unit Subpart I—21CFR 211.160: Laboratory Controls).

—Procedures shall be established and maintained for making changes to a process. Any such change shall be verified or validated to ensure that the change does not create an adverse impact.

—In order to evaluate, approve and implement these changes properly, a company should have an effective change management system (from ICHQ10: Section 3.2.2 Change Management).

DEDICATION

This book is dedicated to all Quality professionals in the pharmaceutical, medical device, biotech, and tissue industries who design, manufacture and market Quality products that improve our lives on a daily basis. I have written this book to make your work a little easier. Keep up the fight for Quality.

David N. Muchemu
CEO and Cofounder, CGMP UNIVERSITY INC.
San Jose, California
Winter 2012.

CONTENTS

CHAPTER ONE: -----------------What is a change in a GMP environment?

A Change

Before one talks about change control or change management, one needs to define what a change is from a compliance point of view. In the pharmaceutical, biologic, issues and medical device industries, a change is defined as any of the following four things:

(a) A solution to an existing problem or nonconformity

This kind of change is usually initiated through a change request from a Corrective Action and Preventive Action (CAPA system), or internal Audit. It is aimed at solving a problem caused by an existing nonconformity. Nonconformities usually result from a product or process not meeting its specifications. This is corrected through root-cause analysis to determine the reason for the nonconformity. Then a solution to the nonconformity is proposed via a change request. The proposed change or fix must be validated before it is implemented.

(b) An enhancement

An enhancement is a change initiated to improve the Quality or performance of the following:

- A marketed product
- A validated process, or test method
- A released document
- An existing manufacturing record, or Quality record
- A qualified piece of equipment or material
- The manufacturing environment
- The Quality system: Infrastructure, documentation

(A) An introduction

Any time an assignable source of variation is introduced in the Quality system a change is set in motion with ***the possibility of unintended consequences***. Candidates for this type of change include:

- .A new process
- A new piece of equipment
- New manufacturing material
- A new document
- A new support system
- A new method
- A new manufacturing site and plant
- Qualification of a new supplier
- New technology or technology transfer
- New design and specification

The key elements that must be considered in all introductions are qualification, verification, testing for equivalency, or validation. .At times a combination of these elements is required.

(C) Discontinuation, retirement, or decommissioning of:

A marketed product

- A manufacturing process used in a marketed product
- A standard operating procedure, or Manufacturing Process Instructions (MPI)
- Technology used to manufacture and distribute a marketed product
- Material used in a marketed product
- A Quality record of a marketed product
- A study on a marketed product
- Equipment used to manufacture a marketed product
- A computer system used in the Quality system
- A qualified supplier on the Approved Supplier List (ASL)

Examples:

I. MEDICAL DEVICE

For medical devices the enhancements are usually associated with a change to a design of a marketed device and the release of new drawing revisions. This type of change is accompanied by a change in the Design Master Record (DMR).

It includes:

- Device performance and functional changes
- Device material changes
- Component changes
- Labeling changes
- Packaging changes

These changes are usually related to the form, fit, and function of the device.

II. PHARMACEUTICAL

(a) Finished drugs

For finished drugs, an enhancement involves changes to the master formulae, batch record, and/or regulatory reporting. These are changes to a marketed drug product, which affect the following critical finished product attributes:

1. Safety
2. Quality
3. Efficacy
4. Chemical and physical properties
5. Potency
6. Identity
7. Purity

Such changes must be recorded and reported to regulatory bodies.

(b) Active Pharmaceutical Ingredients

For active pharmaceutical ingredients, enhancements include:

1. Reprocessing an intermediate, or API
2. Reworking batches

Reworked batches are usually accompanied by validation of the rework procedure, testing, and stability testing.

CHAPTER TWO-----------------------------What s Change Control?

The law of unintended consequences states that all actions have effects that are unanticipated or unintended. Thus, a change in one factor may have unintended consequences elsewhere in the Quality system. Change control assesses and manages the risks associated with the unintended consequence of proposed changes. It is a Quality risk management subsystem.

The idea behind change management, or change control, is to evaluate the change request and approve only those requests that meet the threshold for residual risk and impact that the organization as a system can absorb. The unintended consequences or risk are evaluated from several perspectives:

- The risk to the end-user of the product: patient, nurse, and doctor
- The risk to manufacturing processes downstream
- The risk to the Quality of the product
- The risk to the company's regulatory and compliance standing
- Financial risk for the organization
- The risk to the organization's brand
- Business risk: loss of market share, etc.

The phrase *change control* is an oxymoron. A change cannot be controlled; otherwise, it wouldn't be a change. However, the consequences or risks associated with the change and its impact on the organization can be managed so that the impact of the change and the risk it poses is acceptable to the process, product, the organization and the patient.

From the regulated industry point of view, change control is a systematic approach for:

(1) Mitigating risks and the impact associated with a proposed change to protect the organization against product liability, waste, and regulatory exposure.

(2) Protecting the end user of the product or patient against harm and unintended consequences. From a business point of view, a change control is a risk management tool that protects the organization's regulatory standing, product liability, and waste in the form of recalls and customer returns, not to mention rejects. From a patient and doctor point of view, a change control is a risk management process for mitigating risks associated with unintended consequences resulting from changes to the product and process. A high level change life cycle is shown below:

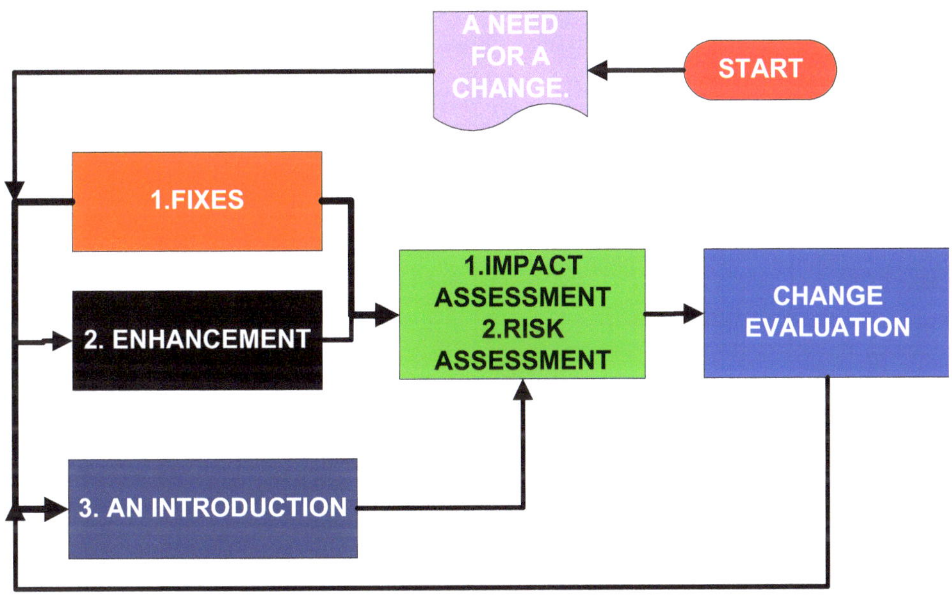

All change requests forwarded to change control must be accompanied by risk management and impact analysis data.

Whether implementing a fix or an enhancement or introducing a new product, process, or equipment into an organization, a change takes place in the organization that takes the organization from a known state to an unknown state. Thus, the impact of the change and its related risks must be evaluated before implementation.

CHANGES TO A PRODUCT

I. DEVICES

Fixes to existing products and fixes to new products during new product development can affect key elements that contribute to the quality of the product. For medical devices, three key elements determine the level of effort that should be involved in the change control processes for products. The three key quality attributes that determine the level of impact assessment and risk assessment are shown below:

Medical device:

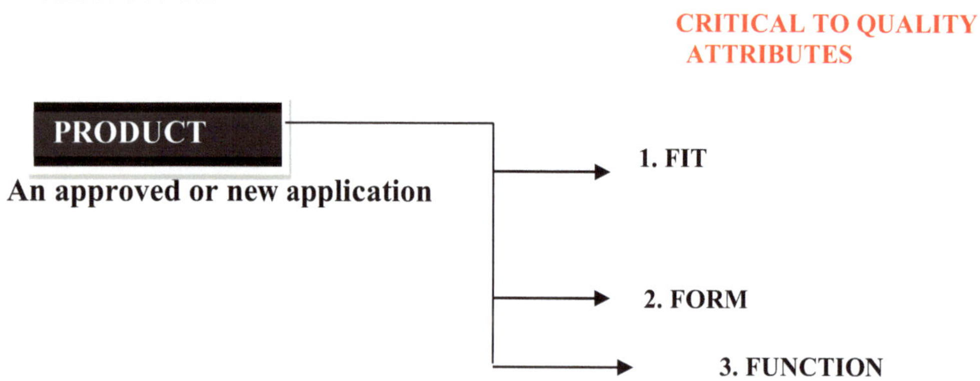

Fit: External dimensions and tolerances
Form: Methods of use, color, and surface finish
Function: Quality, reliability, performance, and set performance limits

Critical to Quality attributes that determine the level of risk management required

Any change to fit, form, and function is accompanied by potential risks that can affect the device, leading to long-term effects that may only surface years later with the threat for product recalls and product liabilities, not to mention regulatory fines.

The scope for product changes includes but is not limited to:

- The design of the device
- Device components
- Software
- Production equipment
- Manufacturing material
- Manufacturing process
- Quality system procedure
- The device aster record
- Design history File
- Quality records
- Device specifications

It should be noted that the practice of Good Documentation Practices and records management is crucial in change control. Effects of unintended consequences have a way of showing up in the Quality system long after the change has been implemented and closed.

II. DRUGS AND BIOLOGICS

The four product Quality attributes that must be evaluated for any drug and biologic change request are shown below:

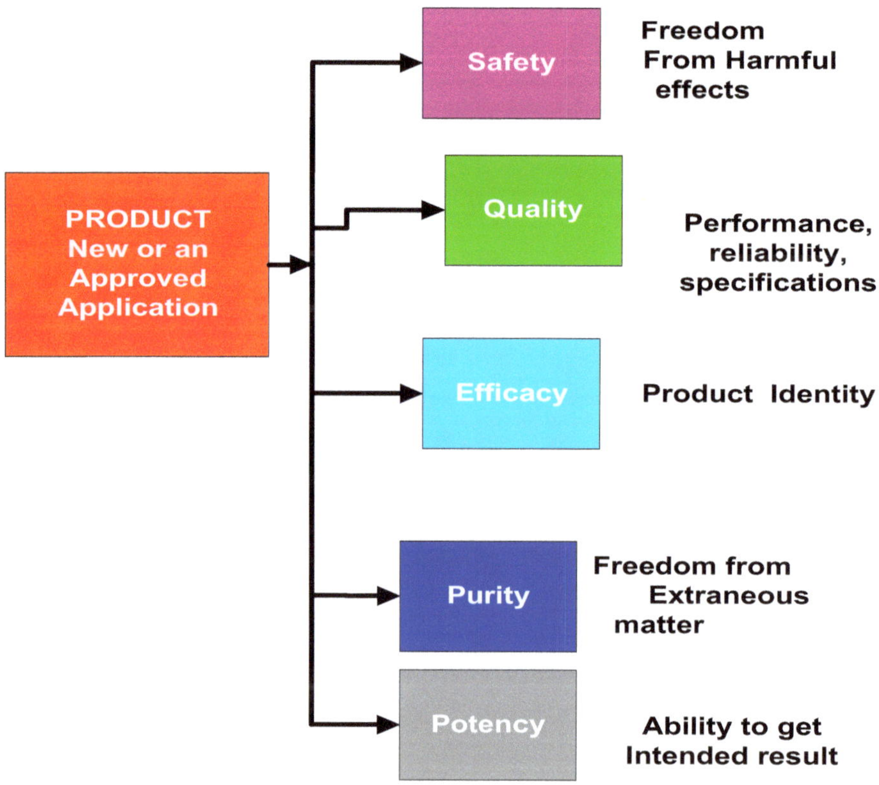

From a quality and regulatory standpoint, changes to these product quality attributes have the potential to expose an organization to liability and increased cost through scrap, in-process rejects, customer returns, product recalls, and rework. These changes can also lead to regulatory compliance issues. It is therefore important that change requests associated with these quality attributes be handled through a formal change control quality subsystem and approved by a quality unit. These change requests should only be approved based on solid scientific knowledge and proven quality improvement methodologies. The aim should be to protect the end user of the product—the patient—while improving the quality of the product and the bottom line.

CHANGE CONTROL

Change control starts with a change request:

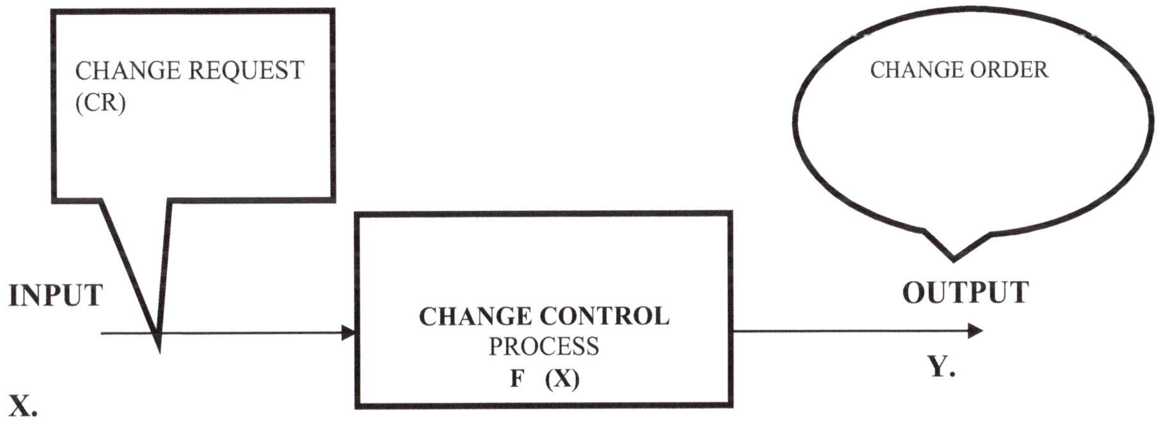

Process: Y= F (X) = Change order

The input of a change control process is a change request. After evaluation, a change request is either:

1. Approved as is
2. Rejected
3. Approved with amendments
4. Deferred: more information required

The output of a change control process is a change order for an approved change request.

Changes during Product Development:

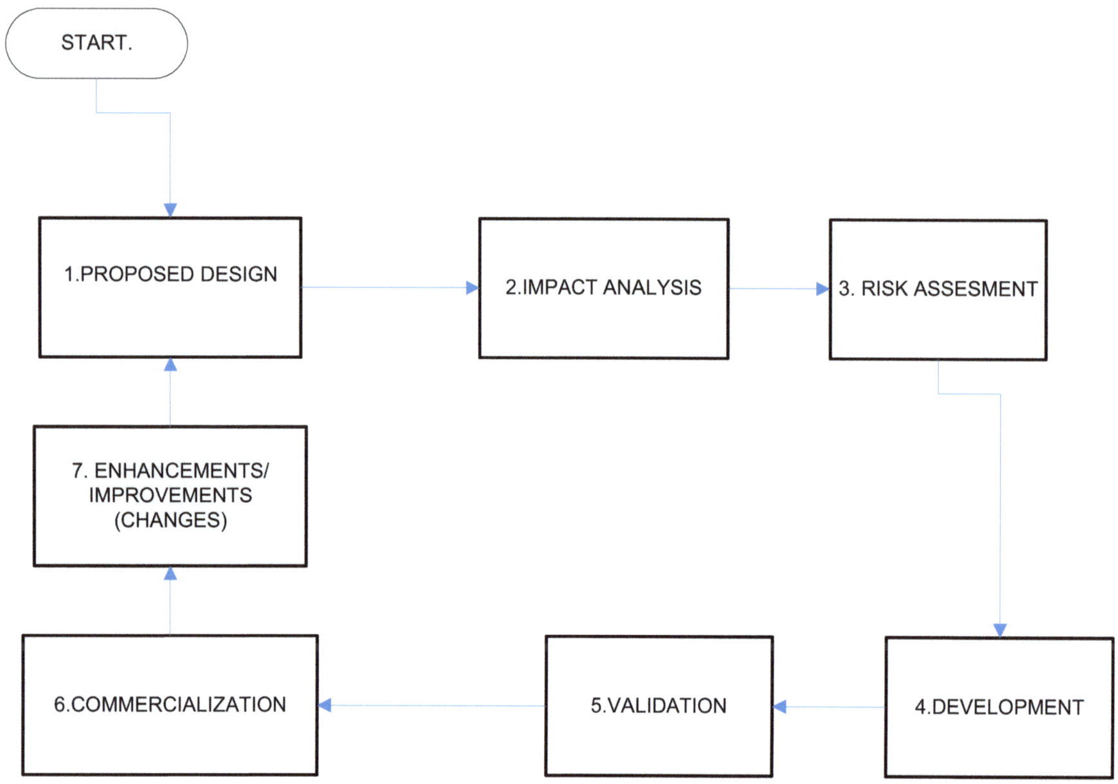

Changes to a new design have to be controlled through the formal change control system regardless of the industry. Once enhancements to the new design are proposed, the impact of the proposed enhancements has to be looked at by performing impact analysis. Impact analysis is usually expressed in a monetary value. For the United States, impact analysis is measured in dollars and cents. The other parameter for impact analysis is return on investment, or cost-to-benefit ratio. Simply put, the cost-to-benefit ratio is computed by dividing the investment in the change initiative by the benefits of the investments. Frankly put: some changes are not worth the cost in resources.

Post-market product Changes:

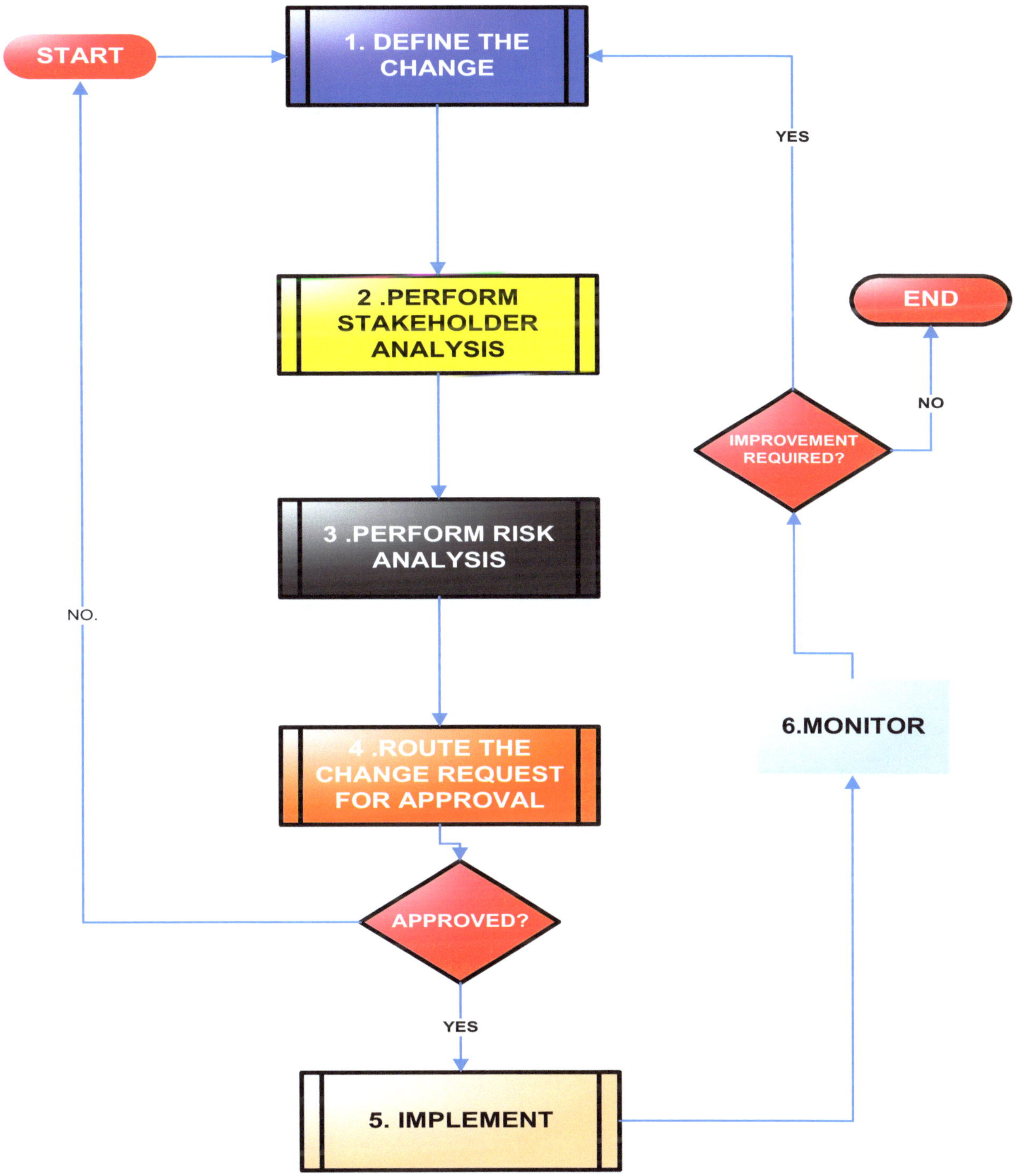

Post-approval product changes initiated to fix issues discovered post-approval should go through hazard analysis (risk management) and be validated. In addition, the impact of proposed changes should be looked at prior to implementation.

CHAPTER THREE: ---------------------------------Change Control Life Cycle

Most change control processes have five phases. Each phase comprises a series of steps that add value to an input, or change request. This results in an output, in the form of either an approved change or a rejected change request. The type of change control process you have should be driven by the culture of your organization, the infrastructure, and the complexity of the product your company makes. As a rule of thumb, a change control system should cover the following phases:

- Change initiation
- Change evaluation
- Change implementation
- Effectiveness check
- Closure

Each phase adds value to an input to produce an end product or an output for the next phase. There are several internal customers to change control processes. A change control manager is a customer for the initiation process, whose output is a completed change request form. Management with executive responsibility or a designee is a customer of the overall change control system. His or her end product is a list of all approved changes, capacity management data, and implementation timelines and schedules. A high level change control life cycle for moderate and major changes is shown below:

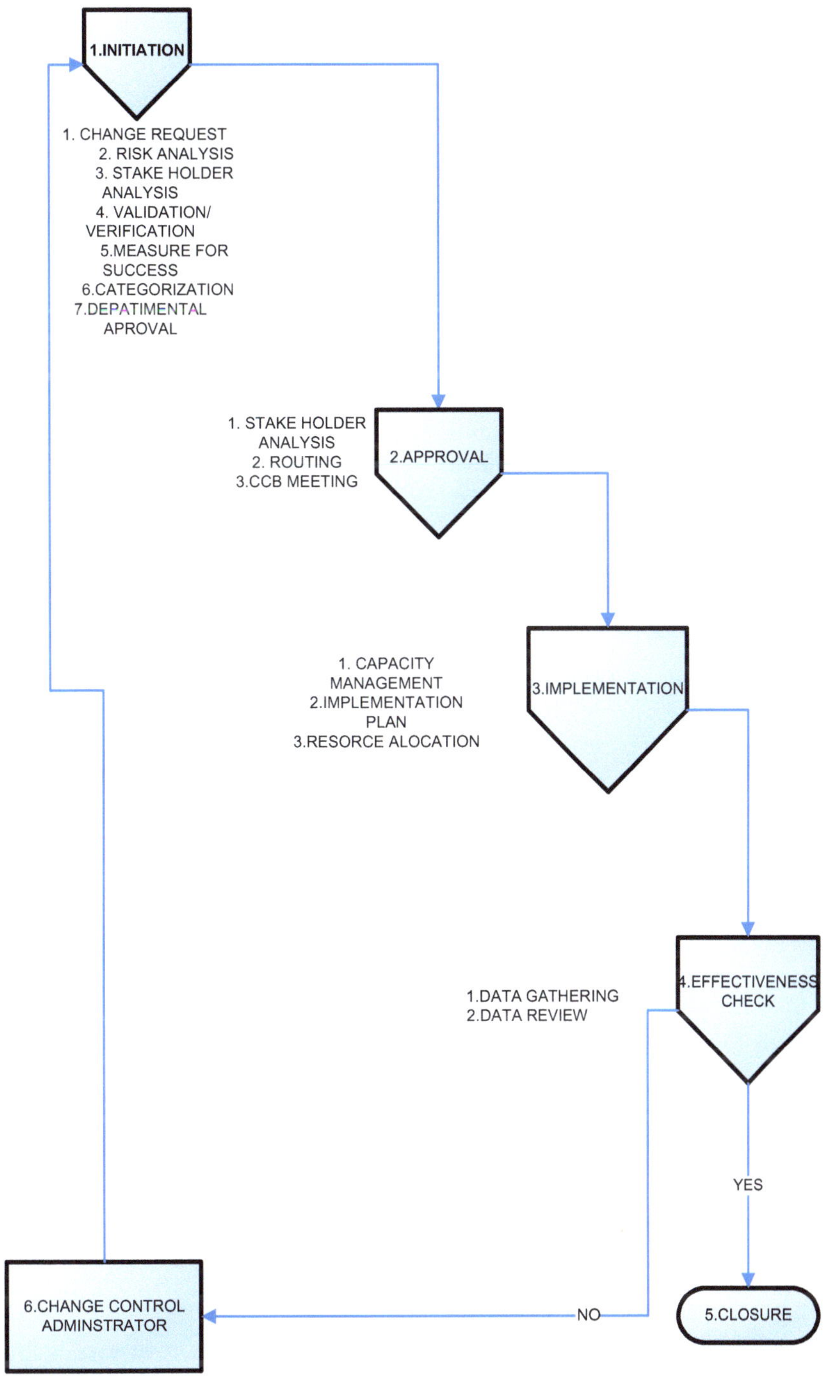

Change control life cycle for a moderate change request

PHASE 1: Change initiation starts with completion of a change request (CR), followed by risk analysis, stakeholder analysis, validation, or verification followed by departmental approval.

PHASE 2: Approval starts with stakeholder analysis and routing, followed by routing and CCB meeting.

PHASE 3: Implementation starts with capacity management, implementation planning, and resource allocation.

PHASE 4: Effectiveness-check starts with data gathering followed by data analysis.

CLOSURE PHASE: Takes place when the data gathered supports the criteria for success set by the initiator. If not, it is communicated to the initiator by quality/change control coordinator.

PHASE DELIVERABLES

Each change control phase has deliverables that must be completed before the next phase of the life cycle is started. The table below is a summary of the change control phases, process, process steps, and expected outputs per phase:

PHASE	PROCESS	PROCESS STEP	OUTPUT
1. Initiation	• Change request		CR, and CR #
	• Risk analysis		Risk data
	• Impact analysis		Expected cost and returns
	• Stakeholder analysis		List of people or functions to be affected
	• Validation/verification	• Define measure of success	Acceptance criteria
		• Change categorization	Change category
		• Departmental approval	Signature
2. Evaluation	• Routing		Complete package
	• Stakeholder input		Analysis report
	• CCB meeting		ECO, rejected CR
3. Implementation	• Planning		Implementation plan
	• Capacity management		Implementation schedule
	• Resource allocation		Ownership
4. Effectiveness check	• Data gathering		Gathering plan: who, what, when
	• Data analysis		Data analysis methodology
5. Closure	• Data review		Final report
	• QA approval		Signatures

Phase deliverables

CHAPTER FOUR------------------------------------Change Categories

Change initiatives or requests are classified based on their impact on and risk to the product, process, end user, and the quality system. Change requests with high risk and impact on the product, end user, process, or quality system usually fall into four categories: Minor, moderate, temporary and major changes.

MAJOR CHANGE

Major changes are those that have been determined **through risk analysis** by the initiator to have **higher potential of having undesired effects** on the product, process, patient (end user) and the quality management system. There are several risk analysis tools used in change control. They are discussed in the next chapter. They include:

- Fault Tree Analysis (FTA)
- Hazard Analysis and Critical Control Points (HACCP)
- Ishikawa: Fishbone
- DFMEA
- PFMEA

Most organizations use **PFMEA** (Process, Failure Mode and Effect Analysis), and **DFMEA** (Design Failure Mode and Effect Analysis) as tools for risk analysis in change control. The only distinction between the two is that the acronym DFMEA is used for design-related changes and PFMEA for process-related changes. The methodology is the same.

The RPN, risk priority number, for any major change should be commensurate with the "complexity and criticality of the change request." Major changes require management review after approval. The signature and input of management with executive authority (usually the head of the quality unit) is required prior to implementation. Once validated, major changes may be incorporated in the quality management system. The common risk assessment tool for the regulated industry is specified in ICHQ9. It is covered in Chapter 5 .

MODERATE CHANGE

Moderate changes are those changes that have no known effects on the product, its properties, process, end user, or the quality system. However, these changes need management's oversight and control for potential long-term effect on the organization. Customer notification is usually necessary if products of a particular batch number are affected. Like major changes, moderate changes may require the input and approval of a member of the management team who has executive authority after or before approval. Moderate changes have a medium risk rating and medium impact on products, processes, and quality management systems.

MINOR CHANGE

By definition, minor changes have minimal or no potential effect on products, processes, or the quality management system. They are usually internal (departmental) in nature and require a minimum number of signatures for approval. Some minor changes are a one-time deal. They are referred to as variances. Variances are initiated through variance requests (VR).

Examples of variances requests include:

- Change in work instructions for rework
- Request for use of material of proven equivalency
- Document changes due to typos
- First article components from new suppliers

Minor changes do not require stakeholder analysis. However, they require impact analysis and risk analysis. Change initiatives can also be classified according to the path taken through the approval processes. Minor changes have the shortest route. Moderate and major changes take the same route during approval—the longest route. Examples of the two routes taken by minor, moderate, and major changes during the change life cycle are given below.

KEY PROCESSES FOR A MINOR CHANGE

Minor changes requests usually require only departmental approval and the approval of the Quality unit for approval. A general path for a minor change life cycle is shown below:

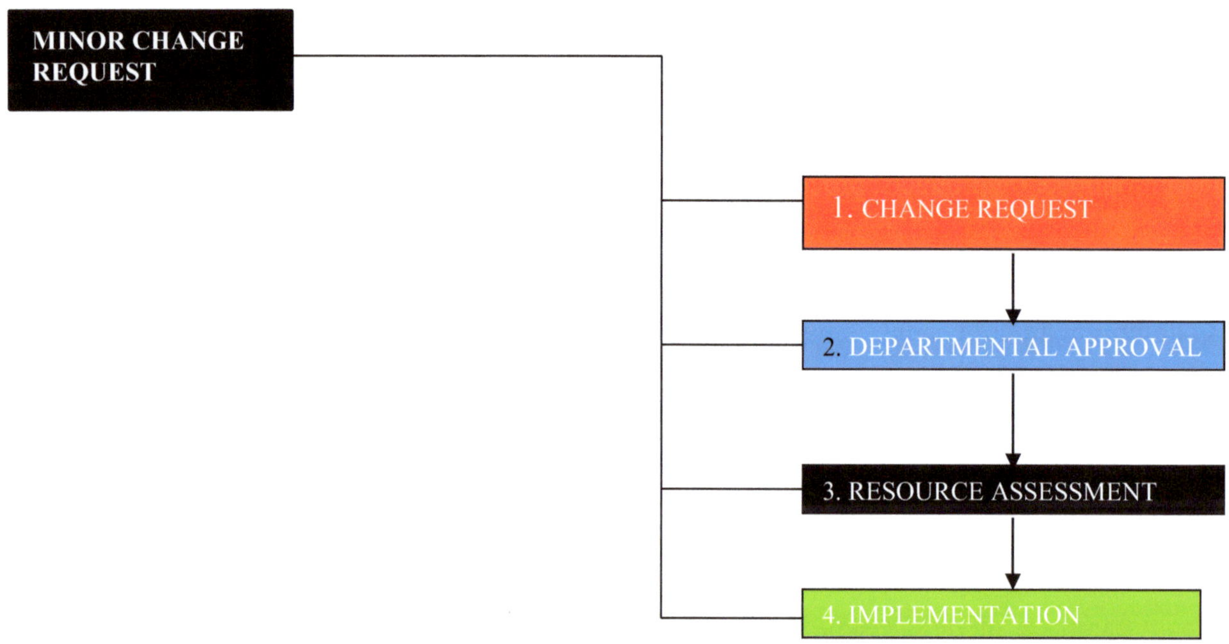

A change request is completed, evaluated, and approved by the departmental managers, assessed for resources allocation, and implemented. Minor changes are mostly departmental and may only require three signatures for approval. A quality representative's signature is required.

MAJOR AND MODERATE CHANGE PROCESSING

Moderate and major change requests require more attention from the initiator. Most of the work is done by the initiator before it is evaluated by the change control committee or board. An example of steps taken is shown below:

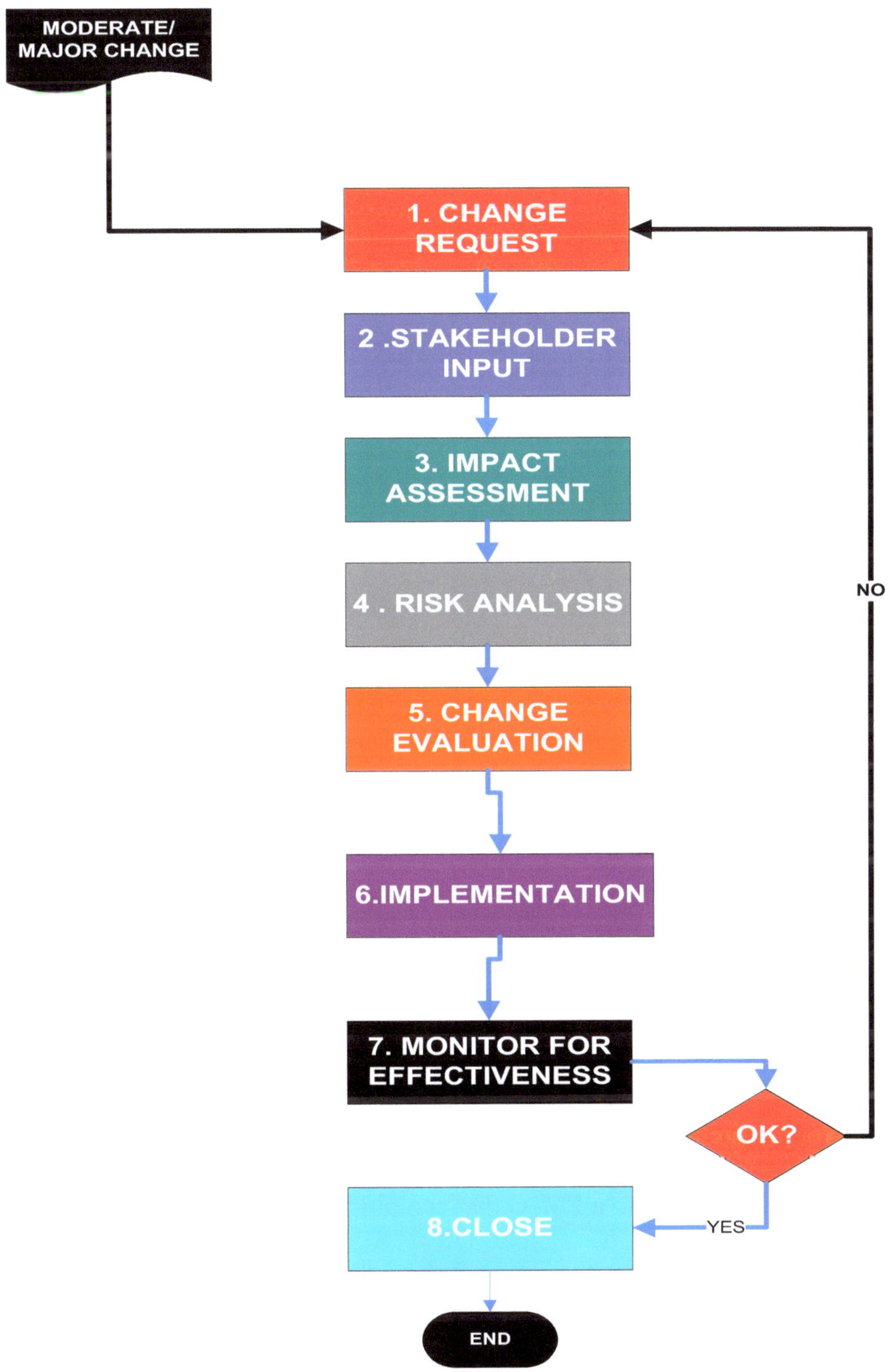

Major and moderate changes go through a minimum of eight processes. The initiator is kept informed by the CCB coordinator of all progress, or lack of it, during each phase. Collection of data that supports the acceptance criteria set for success leads to closure.

REGULATORY CHANGE CLASSIFICATION

Although the QSR and part 211 and 210 are not very clear about how changes should be classified, emphasis is put on pre-market submissions for "changes that could significantly affect safety, or effectiveness and for new, or modified intended uses." For this reason, any change that meets these criteria should be treated as a major or moderate change regardless of whether the product is being introduced into commercial distribution for the first time or is being re-introduced after "modification changes in design, or intended use."

For medical devices, Subpart E 21CFR 807.81 is very clear about what should be considered a "significant change," or a moderate change as we have called it. A moderate change is classified in this subsection as:

- Any change or modification that could affect the safety or effectiveness of the device
- Any modifications in design, material, chemical composition, energy source, or manufacturing process.

A major change, on the other hand, is defined in this section as any modifications in the intended use of the device. This is a very broad definition. By default, some, if not all, moderate changes could be classified as major changes since they have the potential to affect the safety and effectiveness of the finished product.

Other definitions leave a lot to interpretation. Section 21CFR 814.37(c) refers to PMA supplement as "major amendments (changes to a PMA) that contain new data not initially reported to the FDA." These amendments by definition could be referred to as major changes.

Section 21CFR 814.39, on the other hand, calls for a PMA supplement for any changes an organization needs to make after PMA approval if those changes affect the safety or effectiveness of the device. Depending on the vocabulary used in any organization, this requirement can touch on both moderate and major changes. The requirement defines changes that require a PMA (pre-market approval) supplement as:

- New product uses
- Labeling modifications
- Use of a different facility to manufacture, package, or process a product
- Modifications in sterilization methods and/or procedures
- Design modifications
- Performance modifications

These changes can be classified as moderate or major changes, depending on risk analysis data.

TEMPORARY CHANGE

Temporary changes are also called engineering change orders (ECOs), or engineering change Notices (ECNs). They involve implementing processes through a protocol. Actions taken must be returned to their original state after execution. Temporary manufacturing changes are limited to three batches, or three months. If results support the intended output of the temporary change, the ECO or ECN must be converted to a permanent change by opening a normal change request.

CHAPTER FIVE: --------------------------How to perform Risk Assessment

All change requests should be accompanied by three pieces of critical information:

- Stakeholder analysis
- Impact assessment
- Risk analysis

This chapter focuses on the best tools for risk analysis. Risk analysis is achieved by using the ISO 31000:2009 Risk Management model shown below:

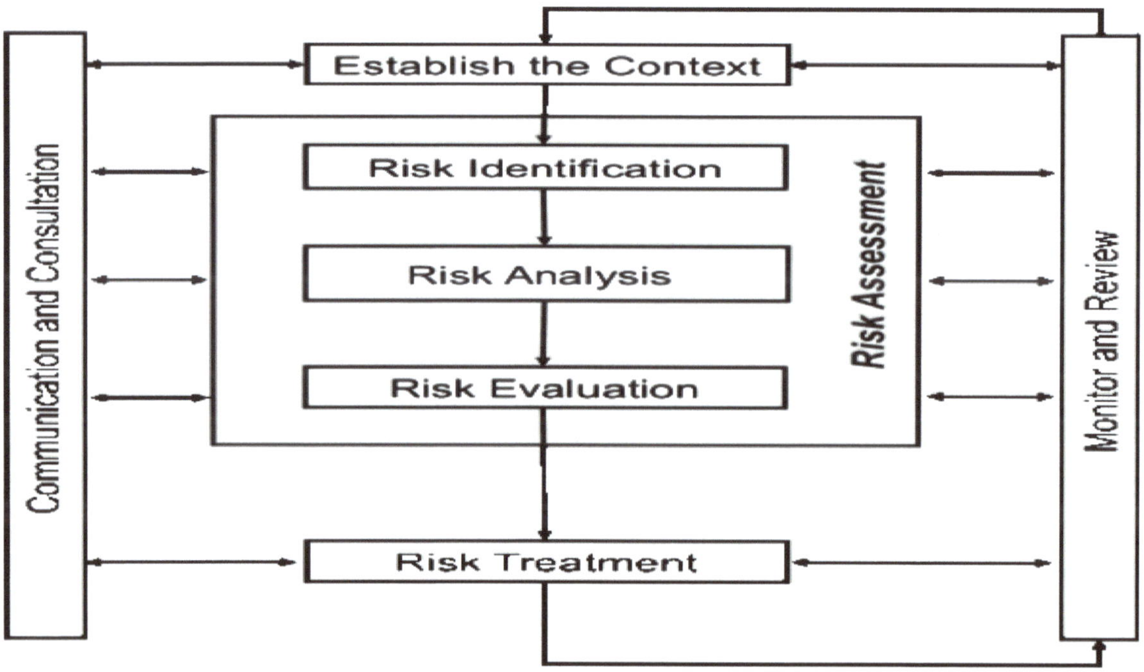

Fig. 1. The risk management process from ISO 31000:2009.

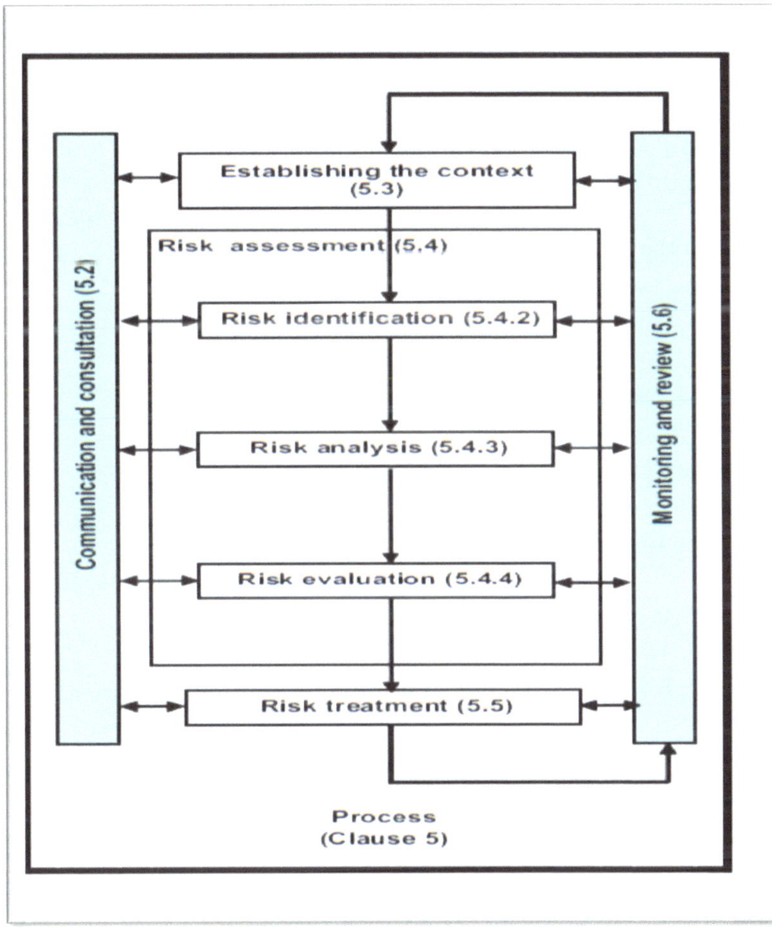

ISO 31000:2009 Risk management model

The model calls for:

5.3 Step one. Establishing the context: Risk in what?

Possible contexts would be:

Changing process parameters

Changing product design

Changing process material

Introducing a new process equipment

Qualifying a new supplier

5.4 Step two: Perform is assessment

5.4.2 Identifying the risk: what risks does what I am trying to do pose?

5.4.3 Analyzing the risk: What are the implications?

5.4.4 Evaluating the risk: how severe and likely is it?

5.5 .5 Risk mitigation/treatment: What should be put in place, or is in place to reduce identified risks?

The ISO 31000:2009 Risk management model is exactly that: a model. It provides the framework through which risk assessment should be performed. It does not tell you what tool to use for risk analysis.

The second risk management model is the ICHQ9. The Quality risk management model is shown below:

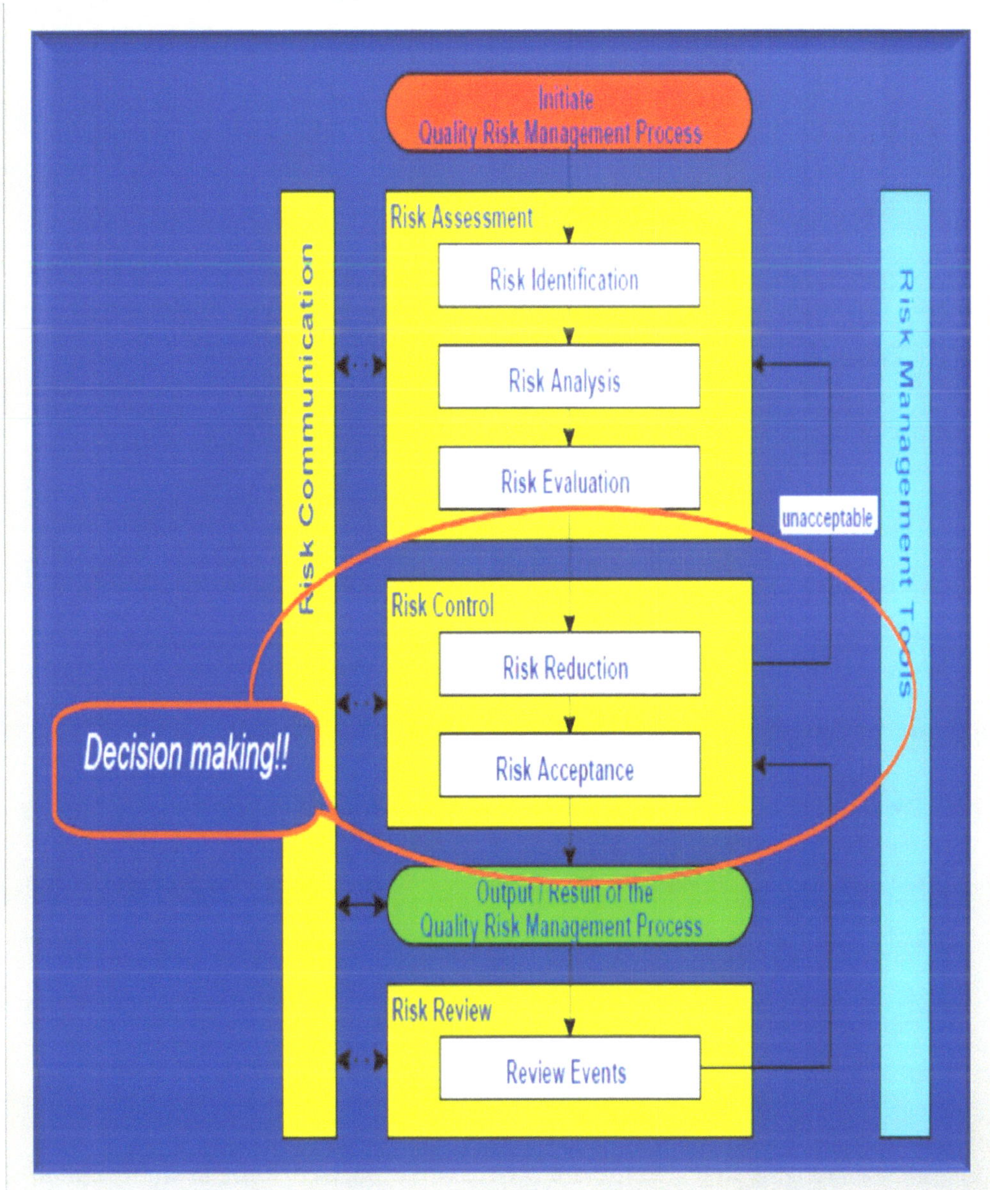

ICHQ9 Quality Risk Management process

Source: http://www.fda.gov

Though a little comprehensive, the ICHQ9 model is based on ISO 31000 risk management principles. The model focuses on:

Phase one: Risk assessment

This phase comprises:

- Risk identification: What risks are posed by what I am proposing?
- Risk analysis: What are the implications?
- Risk evaluation: How severe and common is this?

Phase two: Risk control

This phase comprises the following activities:

- Risk reduction: Mitigation. How can we lower the risk?
- Risk acceptance: Is the residual risk after mitigation acceptable?

Phase three: Risk review

The last step in the analysis is to challenge assumptions made using data:

- Event analysis: Does the data support my assumptions?

The two risk management models are similar. You should choose the one that works for you.

RISK ANALYSIS TOOLS

There are several risk analysis tools. Some are process-related, and some system-related. It is important to use the right tight tool for the right job. The three risk analysis tools used most often are discussed below:

(1) FAULT TREE ANALYSIS (FTA)

Fault tree analysis is a product, process, and system tool. It is based on what are called "AND" and "OR" gates:

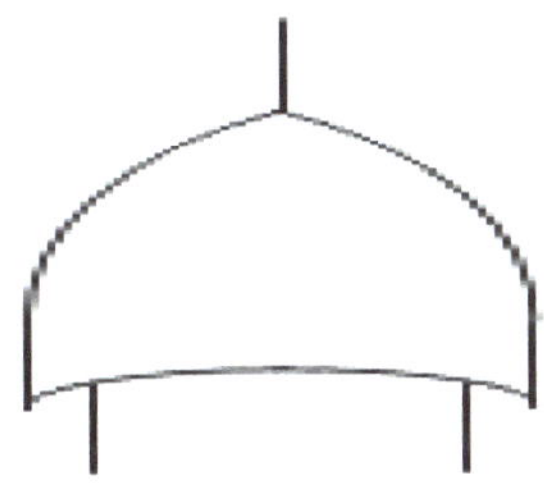

An "OR" gate.

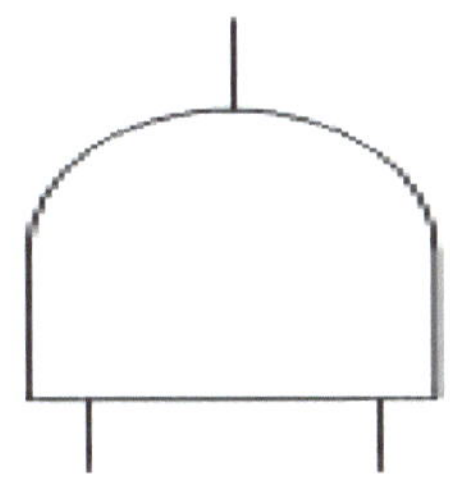

An "AND" gate

"OR" gate: the risk or effect is one of the two conditions

"AND" gate: you need two conditions for the risk or effect to manifest itself

Fault tree diagrams represent the logic relationships between conditions that combine to cause an undesired event, or failure. The event at the top of the fault tree represents the risk of interest and is arrived at by information flowing through logical gates.

Diagram:

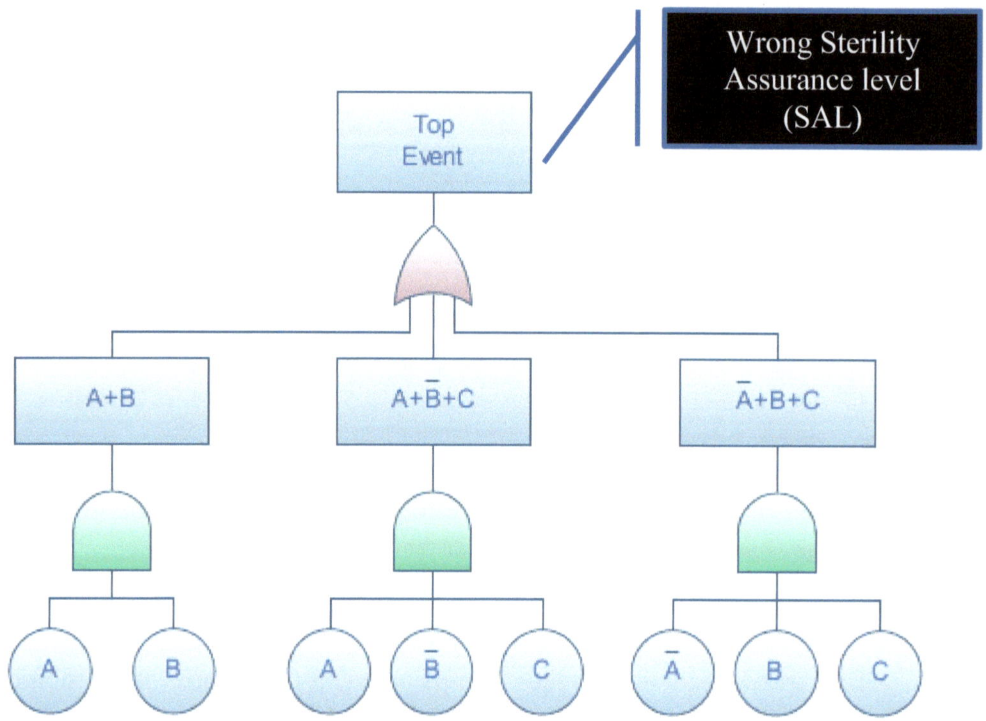

Summary

Bottom layer:

1. Conditions A "AND" B are required to get A+B
2. Condition A "AND" C without B leads to A+C
3. Condition B "AND" C without A leads B+C

Risk: Getting a wrong sterility level for products after a process change. How does my change to the sterilization process contribute to this possibility? How do we mitigate the risk to an acceptable level?

Second layer:

Condition 1, "OR" Condition 2, "OR" condition 3 results in the "Top Event." The top event is the identified risk that needs to mitigated and reduced to an acceptable level for the change request to be approved.

(2) ISHIKAWA: FISHBONE

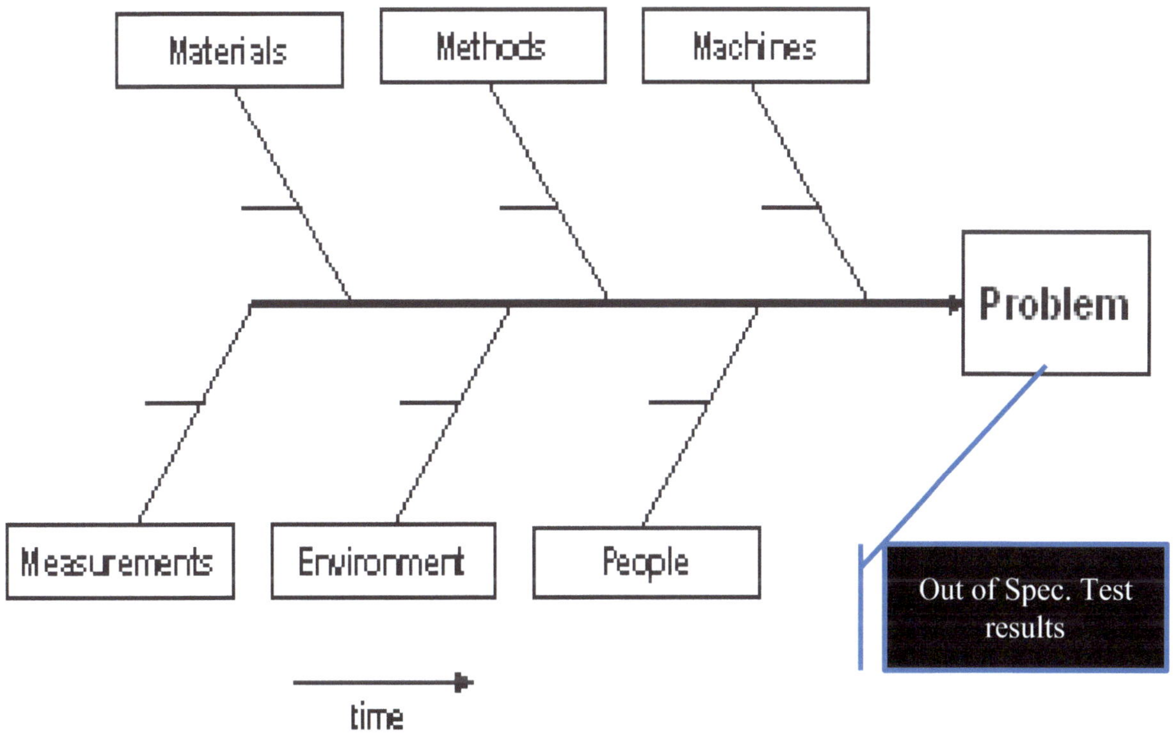

- Ishikawa or fishbone analysis is a risk analysis tool for manufacturing processes. The tool is based on the assumption that all effects, or what you see in a process (risks, events, problems, deviations, etc.) are based on process inputs: Material
- Methods
- Machines

- Man/people
- Mother nature/Environment
- Measurements

The process is the same: the risk has to be defined. In this case, the risk is the "problem." The risk has to be assessed using leading questions. Examples here would be:

- **The risk:** What might go wrong if I change a test method that is in a validated process?
- **Probability rating:** What is the probability of this happening?
- **Severity rating:** What are the consequences of what might happen?

Risks should be identified in reference to the other five process inputs. The final step would be risk mitigation. Risks identified must be lowered to an acceptable level.

Risk mitigation: How can my request to change test parameters lead to this risk? What do we have in place to stop that from happening?

(3) PROCESS FAILURE MODE AND EFFECTS ANALYSIS (PFMEA)

PFMEA (Process Failure Modes and Effects Analysis) is a process risk analysis tool that relies on Quality risk levels to anticipate and prevent process failure at every step. A Quality risk level is a product of three factors: (1) severity of the risk, (2) the likelihood of the risk to occur, and (3) how well the risk detection method in place works. The Quality risk level is measured by a Risk Priority Number (RPN), detailed in the section below.

Risk Priority Number

The risk level for each process step is quantified using an RPN that typically ranges from zero to 1,000, with 1,000 being the highest possible risk level. The RPN is a product of three risk factors, all ranked on a scale of 1 to 10. Theoretically, the highest RPN number for a process step is 1000 (10x10x10), and the lowest is 1 (1x1x1x1).

RPN = Severity x Occurrence x Detection

PFMEA risk factors	
Risk Factor	**Description**
Severity	How severe a given defective condition would be to the customer, process, and Quality system
Occurrence	The best estimate of how often the failure will occur in the process
Detection	The likelihood that the defective condition will be detected prior to reaching the customer or any end user of the output

Severity Scale

The scale is based on the severity of the failure to the customer, process, Quality system and the organization.

PFMEA severity rankings	
Severity Scale	**Description**
10	Hazardous, without warning: Catastrophic, results in death, recalls, regulatory issues
9	Hazardous, with warning: Possible death, possible recalls, regulatory issues
8	Very High: Serious, process failure, regulatory issues, affects the Quality system
7	High: Impacts end-user, process and Quality system
6	Moderate: Impacts end-user, and Quality system
5	Low: Impacts Quality system. Corrective action required
4	Very Low: Impacts process output
3	Minor: Event .A Candidate for preventive action
2	Very Minor: Recordable event
1	None

Severity levels of 9 and 10 are typically reserved for hazardous situations, with a severity of 10 reserved for hazards that the customer is not warned about ahead of time. For example, an electrical shock hazard would typically be rated with a severity of 10 to 1.

Occurrence

Occurrence value represents how often a given failure is likely to happen; robust processes with well-defined steps have a low likelihood of the failure happening. **Occurrence rankings**	
Occurrence Rank	**Likelihood**
10	>100 occurrences per 1,000 process runs likely
9	50 occurrences per 1,000 process runs likely
8	20 occurrences per 1,000 process runs likely
7	10 occurrences per 1,000 process runs likely
6	5 occurrences per 1,000 process runs likely
5	2 occurrences per 1,000 process runs likely
4	1 occurrence per 1,000 process runs likely
3	0.5 occurrences per 1,000 process runs likely
2	0.1 occurrences per 1,000 process runs likely
1	< 0.01occurrences per 1,000 process runs likely

Detection

Detection ranking refers to how well risk detection methods are built into the process to mitigate risk and lower the possibility of not detecting the failure or risk.

Detection rankings	
Ranking	**Description**
10	Absolutely impossible to detect
9	Very remote chance of detection
8	Remote chance of detection
7	Very low chance of detection
6	Low chance of detection
5	Moderate chance of detection
4	Moderately high chance of detection
3	High chance of detection
2	Detection is almost certain
1	Detection is certain: There is no way we cannot catch the failure.

A detection level of 10 implies that it is virtually impossible to detect a given process failure when it occurs. A detection level of 1 implies that built-in methods (tests, inspections, etc.) in the process are absolutely guaranteed to catch the defect when it occurs. The better the detection methods, the lower the ranking.

Failure Mode

The failure mode of a process step refers to what could go wrong, or the way the failure occurs.

Effect

The effect of the failure refers to the consequence of the failure, or the way the failure manifests itself.

PFMEA TABLE

An example of a PFMEA risk analysis table is shown below:

PFMEA																CHANGE CONTROL#:
PROCESS OWNER:								FMEA No/Rev: 0000/01								
Date: XX/XX/XXXX								PROCESS NAME:								
												Action Results				
Process step.	Function	Potential Failure Mode	Potential Effects or Failure	Svrty	Potential Causes of Failure	Ocrnc	Current Controls for Prevention/ Detection	Dtctn	RPN	Risk Mitigation	Ownership	Action Taken	Svrty	Ocrnc	Dtctn	RPN
1																
2																
3																
4																
5																
6																
7																
8																
9																
10																
11																
12																
13																
14																
15																
16																
17																
18																
19																
20																

PFMEA risk analysis table

Risk is assessed on three fronts:

- Risk to the end-user of the product--the patient and doctor
- Risk to compliance with regulations, and registrations
- Risk to the business

  The last bullet is especially important because the effects of a change may take long to show up in the Quality system as recalls, customer returns, and—worst-case scenario—the death of a customer leading to a class action suit. Risk analysis should be given the seriousness it deserves.

Recommended tools include the risk management cycle below:

RISK MANAGEMENT CYCLE

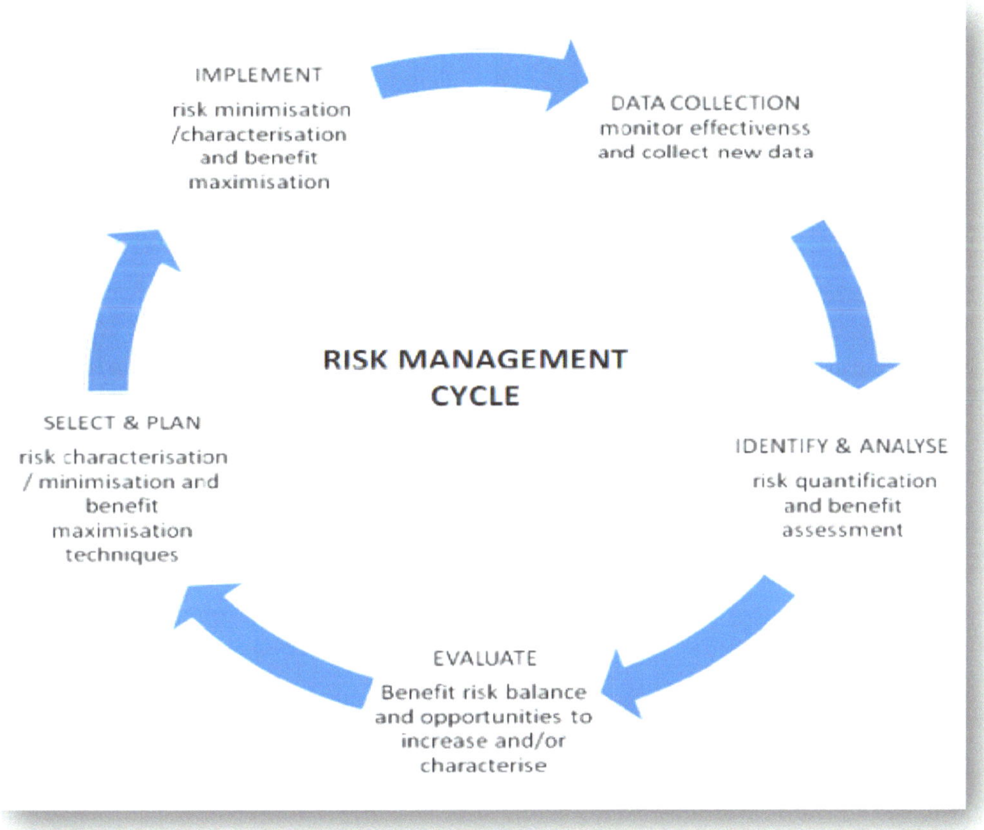

Source: EMA's Guide to Good Pharmacovigilance Practices, module five

This risk management model can be used to evaluate the risk associated with the proposed change from several levels:

- Risk to the end-user of the product--patient, or doctor
- Risk to processes upstream and downstream
- Risk to the Quality system
- Risk to compliance, or regulatory standing
- risk to the business, including waste, branding and litigation

For any change request to be approved through a change control system, the risk associated with it on all levels must be minimized and its benefits maximized.

CHAPTER SIX: -------How to perform Impact Assessment and stakeholder Analysis

Any major and moderate change request must be accompanied by three pieces of information:

- Stakeholder analysis
- Impact assessment
- Risk assessment

Stakeholder Analysis

Stakeholders are people, organizations, or third parties outside the organization that may have a vested interest in the change request. Stakeholders may include:

- Regulatory bodies
- The Quality unit
- The validation group
- External suppliers
- Internal suppliers for a process
- External customers—the patient, and doctor

The leading question for stakeholder analysis is: who might have a vested interest in the change I am proposing, and how do I involve them?

Once the question is answered, a stakeholder matrix is put together, and their interest in the change request is identified. An example of a stakeholder analysis matrix is shown below:

STAKEHOLDER	INVOLVEMENT	LEVEL OF INVOLVEMENT	OWNERSHIP
1.Quality	Inform	Advisor	Review and Approval
2.Operations	Inform, execute	Resource allocation	Approval
3.Regulatory Affairs	Inform, advise	Compliance resource	Review and Approval
4.External supplier	Inform, production	Manufacturing	Approval
5.Packaging	Inform	Packaging	Design

Stakeholder analysis matrix

The stakeholder analysis matrix provides a map that provides the following information:

- **Who is involved.** Who should be involved?
- **Their level of involvement.** What should their role be?
- **Ownership.** What are they responsible for?

The idea is to talk to people or departments that may have a vested interest in the change you are proposing, to have their input and buy-in. Change does not take place in a vacuum.

Stakeholder Input

Every change request has people and organizations with vested interest in it. Leaders of departments that may be impacted by the change request must provide their input. This is imperative. Stakeholders can either be internal or external. External customers are stakeholders and should be involved in any change request if the change affects either the product or the processes used to produce it. Stakeholder input should be received before the change request is routed for evaluation.

Impact Assessment

Impact assessment addresses **four** questions:

1. Who or what is affected by this change?
2. How are they affected by this change?
3. What needs to be done to make this happen?
4. Why are we doing this?

Let us look at each of these questions in depth:

(1) Who or what is affected by this change?

A proposed change has the potential to cause ripple effects in the Quality system. For that reason, work should be performed up front by the change request owner to establish how the requested change may potentially affect:

- Other departments
- Other sites
- Other geographical locations
- Other projects
- Open CAPAs
- Other changes
- Ongoing projects

The implications of a requested change for these elements must be established and all prerequisites met before the request is presented to change control for evaluation.

(2) How are they affected?

This question addresses the need for prerequisites and declaration of deliverables. It also seeks to address the current and future state of affected parties. The change request owner must ensure that all parties affected are informed about their roles, tasks, and deliverables required to advance from the current state to the future state. Emphasis should be on cost and benefit.

(3) What needs to be done to make this happen?

Each change has prerequisites that must be identified and fulfilled before the change can be executed. The prerequisites must be specified in the change request. For example, if the change request is about qualifying a new component supplier, the new material must be qualified through engineering studies before the supplier can be qualified and placed on the approved supplier list. The prerequisite for supplier qualification is material qualification. If the change request is to optimize a process by changing process parameters, a Design of Experiment study (DOE) is a prerequisite for a change in process parameters. These first three questions are best answered using an impact assessment form. An example is shown below:

WHO IS AFFECTED	WHAT IS AFFECTED	HOW AFFECTED	CURRENT STATE	FUTURE STATE	PREREQUISITES
1. Quality	CAPA #00143	The change is the solution to the nonconformity	The Open CAPA	Closed CAPA	1.Investigation 2.Validation/ verification
2. Operations	CAPA#00143	Implementation	Pending	Implemented solution	Validation/ verification
3. SFO site	Production	Material inventory	1. Supplier going out of business 2. Low material inventory	1. New approved supplier On ASL 2. Sufficient levels of material inventory	1. Material Qualification. 2. Supplier Qualification 3. Supplier Audit

Impact assessment form (i)

Your impact assessment should address the following:

- Current state and future state
- Linkages in the Quality system (projects, CAPA, other changes, etc.)
- The business case for the change requested

An impact assessment is a prerequisite for a change request evaluation.

(4) Why are we doing this?

The change requestor has to make a business case for why he or she is requesting a change from the current state to the future state.

Examples of questions that have to be asked would include:

- Why are we qualifying a new supplier?
- Why are we changing process parameters for a marketed product?
- How is what you are proposing to do better than what we are doing now?
- What is the cost of doing nothing?

Change costs money and man hours. That cost has to be justified. The focus from a business standpoint should be return on investment, or ROI. The final question after impact assessment should be: knowing what we know, is this really worth doing? Not every change requested is good for the organization.

(II) **Impact Assessment Template Form**

The template shown below is the most popular tool used for impact assessment. The impact of the change is assessed based on:

- The product
- The process
- The Quality system
- Compliance
- The end-user

CHANGE REQUEST	ASSESSMENT QUESTION WHAT	 HOW	 WHERE	 WHO	 WHEN	EFFECT ATTRIBUTE	AFFECTED ELEMENT	RISK TYPE	RISK MITIGATION
						MANF. MATERIAL	A. PROCESS		
						MANF. PROCESS			
						MANF.METHOD(MPI)			
						MANF.EQUIPMENT			
						TEST METHOD			
						MANF.ENVIRONMENT			
						MANF.PERSONNEL			
							B. PRODUCT		
						SPECIFICATIONS/QUALITY			
						INVENTORY MANAGEMENT			
						PRODUCT RELEASE			
						PRODUCT DISTRRIBUTION			
						PRODUCT SHELF LIFE			
							C. QMS		
						COMPUTER SYSTEMS			
						OTHER QUALITY SUBSTEMS			
						BUSINESS PROCESS			
						LABELING			
						MARKETING			
						DISTRIBUTION			
						PERSONNEL			
						STANDAD OPERATING PROCEDURES			
						WORK INSTRUCTIONS			
						CAPA			
						CHANGE CONTROL			
						COMPLAINT SYSTEM			
						PACKAGING			
						DOCUMENT CONTROL			
						FACILITIES/OTHER SITES			
						COMPLIANCE STANDING	D. COMPLIANCE		
						SUBMISSION			
						INDENTED USE	E. CUSTOMER		
						QUALITY			
						EFFICACY			
	What is affected?	How is it affected?	Where is it affected?	Who is affected?	When are they affected?	Quality element?			

Initiator:					
Change type:					
Residual impact:					
Type:	product	process	Quality system	customer	compliance
Acceptance:	Acceptable	Not Acceptable	Remediation required		

The leading questions are:

- How does this change affect--------- (For process: manufacturing material, etc.)
- Where does this change affect------ (For process: manufacturing material, etc.)
- Who does this change affect---------(For process: manufacturing personnel)
- When does this change affect----------(For process: manufacturing)
- What does this change affect---------- (For process: equipment, etc.)

The questions are paired against effect attributes in the five affected categories.

CHAPTER SEVEN:------Regulatory Requirements for Change Control

From a regulatory standpoint, change control is a risk evaluation system whose purpose is to protect the end user of the product against unintended consequences of changes to the product, process, and Quality system. According to the Pharmaceutical Quality Management System, changes should be managed throughout the product life cycle.

THE ROLE OF CHANGE CONTOL IN A PHARMACEUTICAL QUALITY SYSTEM

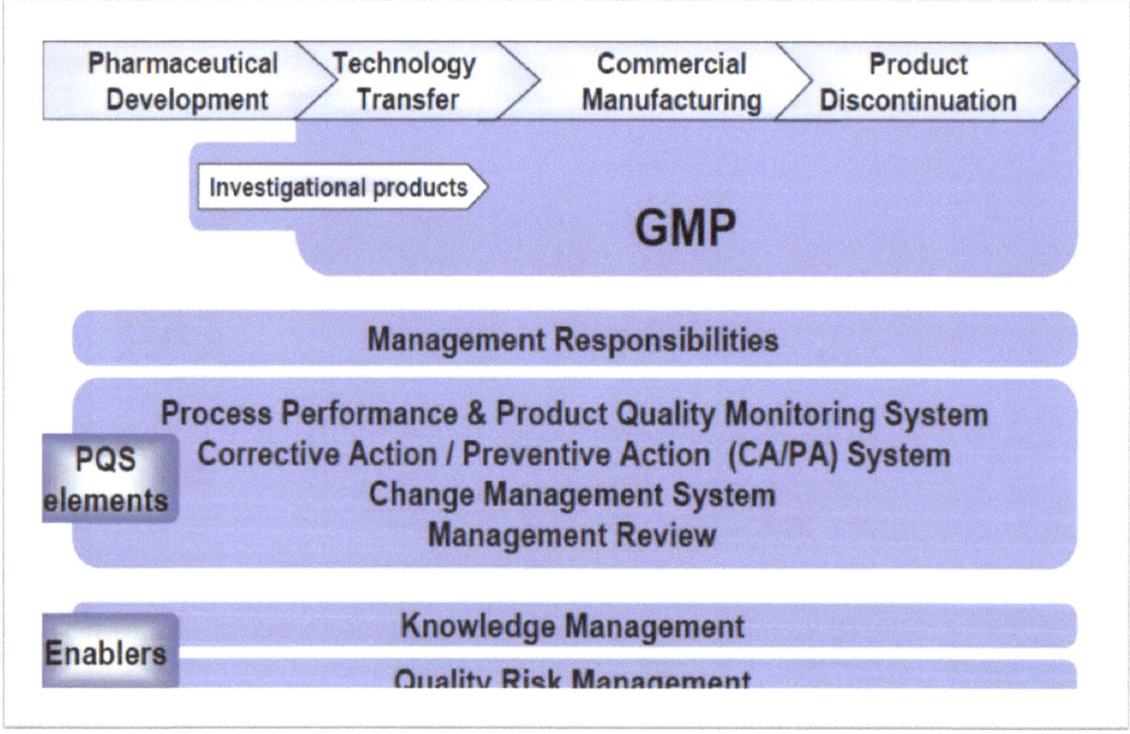

Pharmaceutical Quality System

Source: **ICHQ10**

The Pharmaceutical Quality Management System model advocates change control throughout the product life cycle to drive innovation and continuous improvement. Decisions made on innovation and improvements should be evaluated through a centralized cross-functional change control system. A high level process map for change control should cover the entire product life cycle, as shown below:

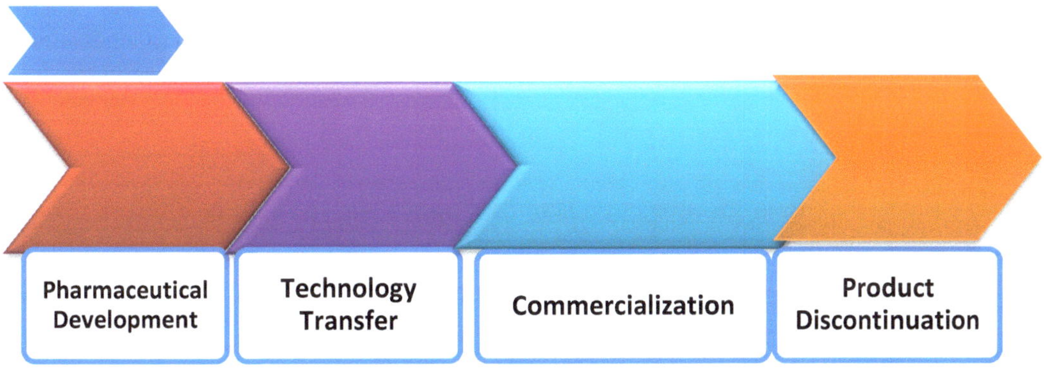

High-level product life cycle change management phases for drugs

Changes from the product life cycle may include:

- Design changes
- Process changes
- Introduction of new technology
- CAPA (Corrective Actions and Preventive Action)
- Changes to validated processes and facilities

  Changes to a product after it is discontinued

REGULATORY REQUIREMENTS FOR DRUGS AND DRUG PRODUCTS

Sections in the Subpart E of Part 211 touch on the requirements for change control of components and drug products:

- 211.68 requires use of **appropriate** deviations when old stock of containers, components, and drug products are used. The key word here is **appropriate** (risk-based).

Subpart F of part 211 touches on changes in production change control:

- 211.100(a) requires that changes in written standard operating procedures or work instructions be drafted, reviewed, and approved by appropriate organizational units (CCB), and they must be reviewed and approved by Quality.
- 211.100(b) requires use of written, justified, and recorded deviations (approved deviations).
- Subpart I touches on laboratory controls:
    - 211.160(a) requires that any changes in laboratory specifications, standards, sampling plans, test procedures, and other laboratory controls be drafted by appropriate organizational units and be reviewed and approved by Quality.
    - 211.160(a) also requires deviations from written specifications, standards, sampling plans, test procedures, or other controls be justified (approved) and recorded.

SUPAC INITIATIVES

Though not legally binding for the pharmaceutical industry, SUPAC initiatives are good guidance for change control. SUPAC categorizes pharmaceutical changes into three categories:

Level 1: Minor Change

- Minimal probability of any risk occurrence
- Unlikely to have any notable impact on quality and performance
- Hardly any detectable adverse effect on quality and performance

Level 2: Moderate Change

- Could have significant impact on quality and performance
- May have a notable impact on functionality
- May have detectable impact on functionality

Level 3: Major Change

- Likely to have significant impact on quality and performance
- Notable impact on quality and performance
- Highly detectable risk to quality and performance

FDA MODERNIZATION ACT OF 1997

Section 116 of this act deals with manufacturing changes for drugs. It calls for holders of NDAs to validate effects of all product-related changes. The act classifies manufacturing changes into three categories:

MAJOR CHANGE

- Prior approval supplements required
- Changes with substantial potential to adversely affect the identity, strength, quality, purity, and potency of the drug
- Changes in quantitative or qualitative formulation or specifications
- Change determined to have substantial potential to adversely affect the safety or effectiveness of the drug

MODERATE CHANGE

- A change with moderate potential to adversely affect the identity, strength, quality, purity, or potency of the drug

MINOR CHANGE

- A change with minimal potential to adversely affect the identity, strength, quality, purity, or potency of the product

A common theme runs through SUPAC, the FDA Act of 1997, and cGMP requirements for change control. Change classification is based on risk to the performance and quality of the product, and ultimately the risk to the end-user or consumer.

Note that the key phrase is: **"potential to have adverse effects"** The question then becomes how do you determine the potential for adverse effects? The answer is something that was mentioned in the previous pages: risk assessment.

MEDICAL DEVICE SATELITE SUBSYSTEMS TO CHANGE CONTROL

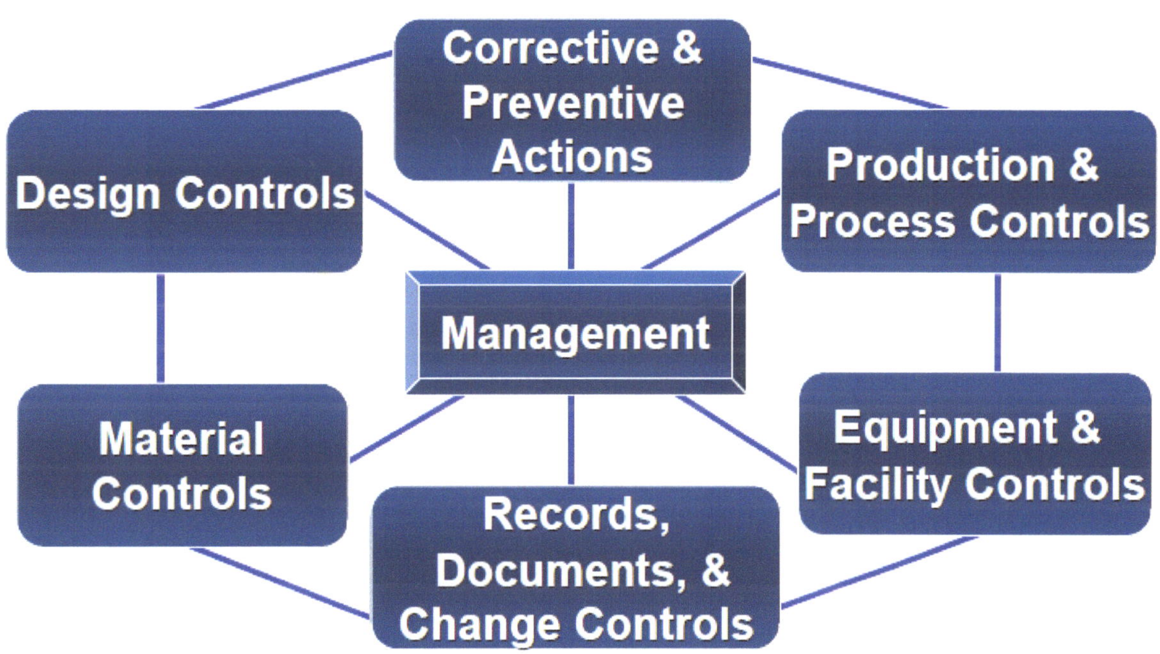

Medical Device Quality System
Source: *Quality System Inspection Technique, QSIT*

The QSIT defines a Quality Management System for medical devices with seven subsystems shown above. The seven subsystems are interrelated and communicate with each other. A change in one can have unintended consequences on other subsystems. Risk assessment and impact assessment is crucial to any change initiation. Each subsystem is a supplier of change requests in the change control subsystem.

SATELITE SUBSYSTEMS THAT FEED CHANGE CONTROL

Medical device changes come from the following Quality subsystems:

A. DESIGN CONTROLS: 21CFR 820.30

This subsystem provides the following changes in:

- Design input
- Design history files (DHF)
- Acceptance criteria
- Product specifications
- Design validation
- Design verification
- Design transfer
- Design review
- Product enhancement

B. MATERIAL CONTROLS

This subsystem comprises the following:

- Subpart E: Purchasing controls
- Subpart F:Identification and traceability
- Subpart I: Nonconforming products
- Subpart M: Records

This quality subsystem produces the following changes:

- Changes in supplier contracts that affect service or material supplied
- Changes in supplier processes that affect material or component supplied
- Changes in component specification
- Change in manufacturing materials
- Material equivalency tests

C. PRODUCTION AND PROCESS CONTROLS

This is one of the most important sources of variation in the quality management system.

Changes from this quality subsystem include:

- Changes from process validation studies
- Changes in sterilization standard operating procedures
- Change in process parameters
- Change in labeling
- Change in product inserts
- Introduction of new technology/methods

Changes in test methods

D. FACILITIES AND EQUIPMENT CONTROLS

Changes from this quality subsystem are associated with environmental controls. They include:

- Changes associated with relocation to a new facility
- Changes in environmental control parameters
- Changes in utilities
- New equipment validation and qualification changes
- Computer software and hardware changes
- Computer software validation changes

E. CORRECTIVE ACTION AND PREVENTIVE ACTION (CAPA)

Most corrective action investigations result in the implementation of new methods, introduction of new procedure, or changes in specifications. Changes from the CAPA system include:

1. Implementing a validated or verified solution to an existing nonconformity
2. Implementing a verified or validated solution to the root cause of an existing deviation

In essence, the output of a CAPA system (a solution to a problem, or potential problem) is an input (change request, CR) of the change control system.

F. DOCUMENT CONTROLS

1. Changes in controlled documents

2. Introduction of new work instructions

3. Introduction of new standard operating procedures

4. Introduction of new company policies

5. Quality record documents: DMR and DHR

This list is by no means carved in stone. The satellite supplier subsystems an organization chooses should be unique to its culture and business practices.

At a minimum the following sections of the medical device cGMP must serve as satellite suppliers to your medical device change control system:

- 820.30—Design controls
- 820.50—Purchasing controls (supplier controls)
- 820.40—Document controls
- 820.70—Production and process controls
- 820.75—Process validation
- 820.100—Corrective Action and Preventive Action (CAPA)
- 820.181—Device master record
- 807.87(g)—Pre-market submission changes
- 820.180—Records

These are satellite suppliers to the change control system.

In addition to changes originating from these quality subsystems, there are other changes that may require use of an established change control system for approval.

All pre-market submissions to the FDA require the use of a cross-functional group for approval, as do modifications to the DHR. The rule of thumb for changes involving medical device products that are already on the market is simple: Use a formal change control system to manage potential commercial and regulatory effects of changes introduced to the device in any one of the following situations:

- Design revalidation
- Change in the device history record (DHR)
- 510(k) submission for a modified device
- Performance enhancement during the life cycle of the device (upgrade)
- Corrective action due to performance failure
- Design changes to an existing product on the market

It is good business practice from a liability point of view and also a regulatory standpoint to support all decisions made during the change process with scientific facts and statistical data.

The review, evaluation, approval, and implementation of any change request generate change control records. At a minimum, those records should:

Identify the entity being changed

Describe the change approved

Identify documents, products, and processes affected by the change

Document the approval date

Document the implementation date

Contain post-implementation evaluation data

Identify the source of the change request (product, process, or quality system)

Identify the initiator and the owner

Identify affected processes and procedures

Document data associated with validation activities prior to implementation of the change

Document impact assessment: who and what is affected, and how

Document risk assessment, risk analysis, risk reduction and risk acceptance criteria

Document any deviations during implementation

QSR REQUIREMENTS FOR CHANGE CONTROL

According to the Quality System Regulations (QSR):

- Approved changes have to be communicated to the appropriate personnel in a timely manner
- Each manufacturer should maintain records of document changes
- Each manufacturer should have established change control procedures including change control forms
- All changes do not need the same degree of evaluation and approval
- Changes should be made according to approved company procedures and policy
- Making uncontrolled changes is a violation of sections 820.30, 820.40, 820.70, 820.75, and 807.81
- Each changed device, accessory, labeling, packaging, and process should be thoroughly verified and/or validated and reviewed by change control board
- Remedial action changes for service/field should be handled through a change control system

- Modification to a device or manufacturing process may require pre-market notification or pre-market approval supplement
- Appropriate and adequate quality assurance checks should be performed after implementation
- The burden for determining whether pre-market notification is required after a change or modification of an existing device lies with the manufacturer
- If the change affects safety or effectiveness, another pre-market notification submission (510k) should be submitted to the FDA
- New or modified intended use changes require 510(k) submission
- Management determines the threshold level for new pre-market notification
- A manufacturer can continue to distribute the unchanged device while waiting for FDA review of the submission
- Pre-market notification submission to the FDA after a major change should be accompanied by risk analysis and validation data
- **Change requests** become APPROVED CHANGES, or **change orders** after approval

PRE-MARKET CHANGES

Pre-market Submission

Regulatory requirements for change control during pre-market submission are specified in section 807.87(g). According to this section, pre-market notification submission should "include appropriate supporting data to show that the manufacturer has considered what consequences and effects the change or modification or use may have on the safety and effectiveness of the device."

CHANGE CONTROL REQUIREMENTS FOR BIOLOGICS

REGULATORY REQUIREMENTS

Subpart C of part 601 touches on change control for approved applications:

- **601.12**: This part requires applicants to inform the FDA of any change in product, production process, quality controls, equipment, facilities, labeling, and responsible personnel in the approved license before distributing any product made using the change.
- **601.12**: Requires demonstration through validation or studies of lack of adverse effects after the change in the identity, strength, quality, purity, or potency of the product.
- **601.12(b):** Requires supplement submissions for all major changes—"Changes with a substantial potential to have an adverse effect on the identity, strength, quality, purity, or potency of the product."
- **601.12(b)(2):** Defines major changes as:
  1. Changes in qualitative or quantitative formulation or other specification in approved applications

2. Changes that require completion of human study to demonstrate equivalency
3. Change in material
4. Changes to sterility methods

- **601.12(b)(3):** Requires the applicant to get approval from the FDA before the product is distributed (after a major change)
- **601.12(c):** Requires submission of supplement for any change in product, production process, quality controls, equipment, facilities, or personnel that has a moderate potential to affect the identity, strength, quality, purity, or potency of the product.
- **601.12 (c) (2):** Defines moderate changes as:

  1. Changes in testing site (movement from one facility to another)
  2. Use of different equipment
  3. Use of new equipment

- **601.12 (5)(d):** Defines minor changes as:

1. Changes that can be described in an annual report
2. Changes in product, production process, facilities, quality control, or responsible personnel with minimal potential to adversely affect the product's identity, strength, quality, purity, or potency.

- **601.12(5)(f):** Requires labeling changes to have supplement submissions approved before the product is distributed

Subpart B of part 610.9 touches on modifications of test methods.

- **610.9:** Requires changes to test methods, manufacturing processes, or parameters be accompanied by validation data to prove assurance equal or greater than the initial level of assurance of the product, process, or test method.

CHANGE CONTROL REQUIREMENTS FOR TISSUE PRODUCTS

Section 1271.225 covers process changes for tissue and cellular therapy. According to this section:

- Process changes should be validated
- Process changes should be verified
- Reporting of approved changes in a timely manner is required
- Risk analysis to ensure that changes do not adversely impact other processes or the product
- Process changes must be approved before being implemented
- All changes implemented must be recorded
- A change record comprises description of the change, signatures, affected documents, approval date, and implementation date
- Approved changes should be communicated to management with authority in a timely period

Section 1271.230 covers process validations. This section requires that:

- In cases where process outputs cannot be verified by inspection and testing, process validation must be performed

- All changes during process validation must be validated, and documented

Section 1271.180 touches on changes in standard operating procedures. Requirements in this section require:

- Review and approval of all document revisions/changes
- Review and approval of all deviations
- Documentation of all deviations

CHAPTER EIGHT------------------------------Change Control documents

Whether you have a full-fledged web-based change control system or a paper-based change control system, you must document your change control system. A robust change control system should follow your Quality system design. At a minimum it should have the following:

- A change control policy
- A change control standard operating procedure (SOP)
- Impact assessment work instructions (WIs)
- Risk assessment Work Instructions (WIs)

A. A CHANGE CONTROL POLICY

The highest document in the change control system is the change control policy. The policy is essential because it sets the direction and spells out the scope of the change control quality subsystem. A policy should have the following sections:

1.0 POLICY

You are not required to have a change control policy by regulatory bodies. However, it is a no-brainer from a Quality Assurance point of view.

Example: It is the policy of Amana Inc. to continuously improve products, processes, and the Quality system by proposing changes, evaluating proposed changes and approving them through a centralized change control system.

2.0 SCOPE

This section establishes boundaries for the quality system subsections involved.

Example: This policy applies to introductions, fixes, enhancements, and discontinuations from the following quality subsystems:

2.1 Product development

2.2 The CAPA system

2.3 Production and process controls/manufacturing

2.4 Document controls

2.5 Product and process validation

2.6 Supplier Quality management

2.7 Facilities and equipment controls

2.8 Laboratory controls

2.9 Material controls

2.10 Management controls

3.0 RESPONSIBILITY

This section establishes ownership.

Example: The vice president of Quality or designee is responsible for enforcing this policy. **B. CHANGE CONTROL STANDARD OPERATING PROCEDURE (SOP)**

This is the second-tier document that defines roles and responsibilities in the change control system. It specifies who does what in the system and is usually written in a passive voice, or a reporter's voice.

Sections of a Standard Operating Procedure

The following are the parts of a standard operating procedure, or SOP:

1.0 Purpose

This section spells out the intended purpose of the procedure

Example: This procedure provides instruction on how work flows during change initiation, change evaluation, change review and approval.

2.0 Responsibility

This section spells out ownership of the document, usually a member of the management team with executive authority to ensure is suitable for its intended purpose

Example: The Director of Quality is responsible for the implementation and review of this standard operating procedure for its intended use.

3.0 Scope

This section establishes boundaries for suppliers in your change control system.

Example:

This standard operating procedure applies to changes in:

3.1 Documents

3.2 Manufacturing

3.3 Materials

3.4 Suppliers

3.5 Computer system

3.6 Facilities and equipment

4.0 Definitions

This section establishes the vocabulary used in the procedure.

Example:

The following terms, phrases and acronyms are used in this procedure:

WORD, PHRASE, OR ACRONYM	DEFINITION
1. Change control	A system used to evaluate, review, and approve proposed changes based on impact and risk associated with the change
2. Change order	A document issued to implement an approved change request
3. Stakeholders	All the people or organizations with vested interest in the proposed change
4. ECO	Engineering change order: A temporary change
5. Change Control Board	A cross-functional team that evaluates all proposed changes
6. CCB	Change control board

4.0 OVERVIEW

This section gives a high-level view of the tasks accomplished by the procedure.

Example:

4.1 Moderate changes are processed through the change control board

4.2 Major changes are controlled through the change control board

4.3 Minor change may only require two signatures for approval, one of which must be the manager of the department in which the change request originates

4.4 Minor changes require at a minimum signatures from the department head and quality

4.5 The change board meets twice a week

4.6 A change request can either be rejected, deferred, accepted, or amended and approved

4.7 The initiator is responsible for addressing issues in the amended change

4.8 Once approved, the initiator and the owner or sponsor of the change have 30 days to implement the change

4.9 Deferred changes and rejected changes can be appealed with supporting data

5.0 Procedure

Provides directions as to who does what in the system.

Example:

5.1 The initiator completes the change control request form

5.2 The initiator performs risk assessment

5.3 The initiator performs impact assessment

5.4 The initiator forwards the change request with supporting documentation to the change control coordinator

6.0 References

This section is reserved for references used in the standard operating procedures—external and internal sources.

Example:

External

6.1 21 CFR 820.30(J)

6.2 21CFR 820.40(b)

6.3 21CFR 820.70

6.4 21 CFR 820.75

Internal

6.5 POL 987: Change control policy

6.6 SOP 897: Change control

6.7 WI 0978: Impact assessment

6.8 WI 712: How to initiate a change request

6.0 Attachments

Provide attachments to be used as job aids.

Example

6.1 Process map

6.2 Change request form

Example 6.1: Process map

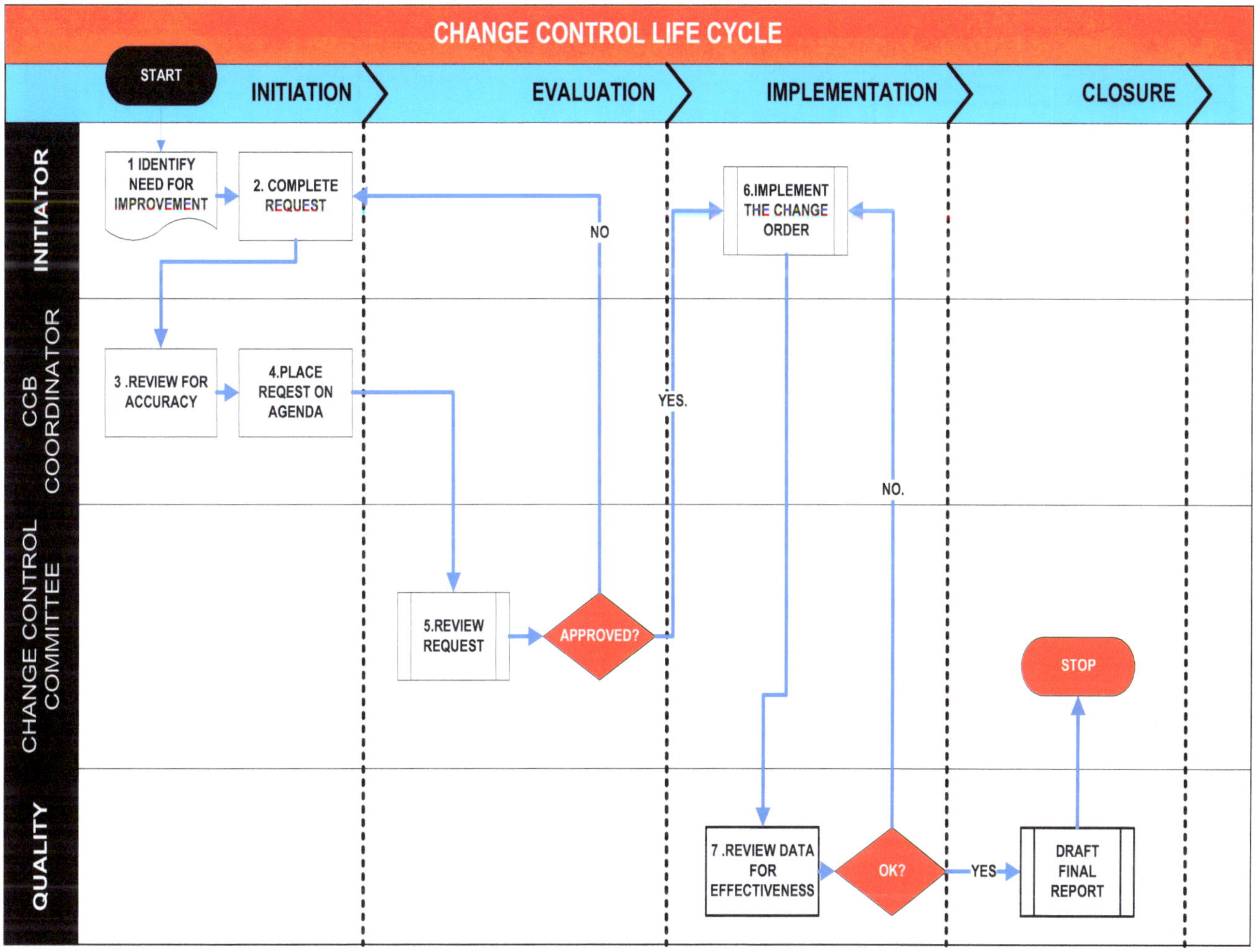

A process map shows how work flows across functions or cross-functional groups. It also establishes ownership. In the above process map, the initiator owns the first task and three processes.

Example 6.2: Change request form

INITIATOR:	CHANGE REQUEST FORM	CR NO.
Dept.:		Date:

A. CHANGE TYPE

☐ A fix
☐ An enhancement
☐ A Solution to a problem
☐ Other(specify):---------------------
☐An Introduction: product, process, equipment, technology
☐ QMS
☐ Temporary/ECO
☐ Variance/Planned deviation

B. CHANGE CLASSIFICATION

☐ Minor
☐ Moderate
☐ Minor

C. CHANGE DESCRIPTION

D. DISPOSITION

☐ APPROVED ☐ REJECTED
☐ AMENDED ☐ DIFFERED

APPROVALS:		Name	Signature	Date
1) Departmental				
2) Manufacturing				
3) Product development				
4) Quality assurance				
5) Regulatory affairs*				

FRM CCR 078 Rev C

C. WORK INSTRUCTIONS

Work instructions are third-tier documents. They provide step-by-step work instructions to perform a task. The number of work instructions in a change control system depends on the number of critical or key processes in the system. The above process map contains process steps and processes. There should be no confusion between processes and process steps. A process has an input and output; a step does not. Three processes stand out on the above process map:

- Completion of a change request
- Review of the change request by change control committee
- Review of post-implementation data for closure

Each of these processes requires a work instruction, or "how to" document. As work instructions explain how to do something, they are always written in an active voice.

Example:

1.0 Purpose

This document provides instructions on how to complete Change Request Form CC 0087 during change initiation.

2.0 Apparatus/Material

2.1 Form CC 0087

2.2 Stakeholder analysis form CC 127

3.0 Responsibility

3.1 It is the responsibility of the change initiator to provide all relevant information regarding the change request.

3.2 It is the responsibility of the initiator to monitor the change request through the approval phase.

4.0 Process Flow:

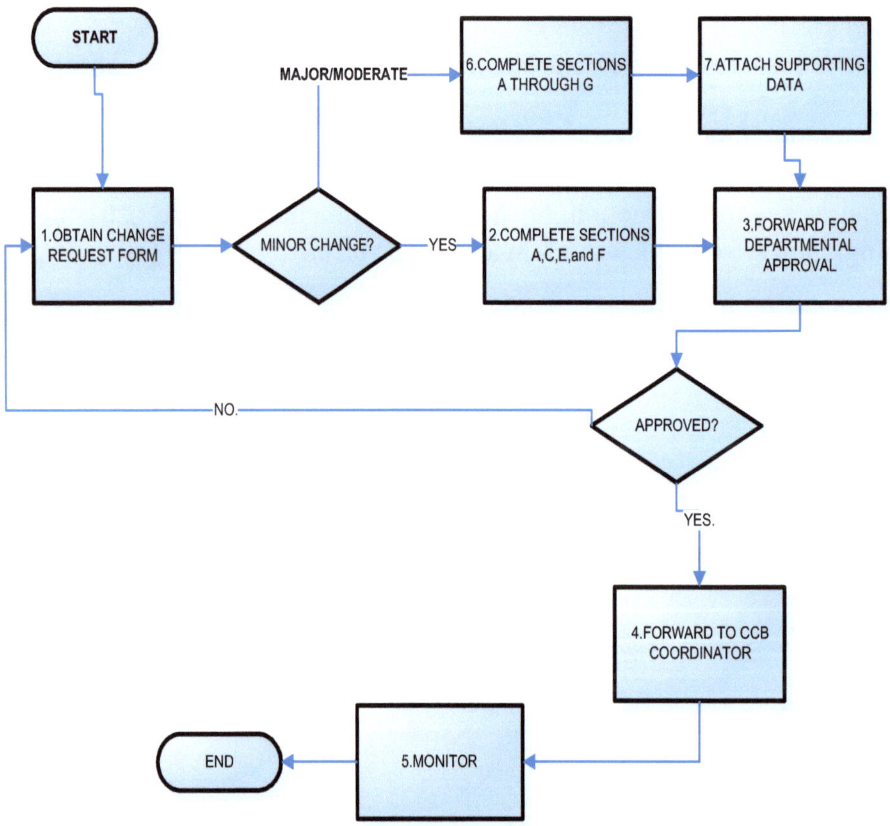

Process flow for change initiation

5.0 Procedure

5.1 Complete the top part of form CC0087 by providing the requested information

5.2 Complete sections A through G for moderate and major changes

5.3 Complete sections A, C, E, and F for minor changes

5.4 Forward the minor change request to the department manager for approval

5.5 Forward the approved minor change request to the CCB coordinator

5.6 Attach supporting evidence to all moderate and major change requests

5.7 Forward the request to the department manager

5.8 Forward approved moderate and major change requests to CCB coordinator

5.9 Monitor the progress of the request

6.0 Records

The following documents should be preserved as proof of tasks accomplished:

6.1 Change request form

6.2 Stakeholder analysis form

6.3 Impact analysis form

Three things must be done for any moderate and major change:

- Stakeholder involvement
- Impact assessment
- Risk assessment

You have to get the buy-in of those affected, assess the impact of what you are proposing to change, and show that the risk associated is acceptable.

CHAPTER NINE: ----------------Part 11 requirements for Change Control

FDA-regulated industries have shifted over the years from a paper-based change control system to computer-based systems. That shift in the industry brought about new challenges for Quality professionals, regulatory affairs professionals and the FDA. At the top of the list were two issues:

- Assurance that electronic signatures were trustworthy and authentic
- Assurance that electronic records were trustworthy and had not been tampered with

Part 11 requirements introduced by the FDA for electronic signatures and electronic records are meant to ensure the above issues are taken into account when a computer system (software, hardware, and firmware) is introduced in the Quality system to manage control documents, especially records. The controls you put in place to maintain the integrity of your electronic records and electronic signatures for your change control system can be system-based (technical), or specified in your procedures (manual).Your controls should address:

- Electronic records that people create
- Electronic signatures that people create and use to sign into the system
- Archived electronic records
- Transmission of electronic records
- Archived and transmitted signatures

For change control, this means you should have controls for signatures and record integrity in place through the entire change control life cycle. Controls should start with access to the change control system.

PART 11 REQUIREMENTS

Section 11.10

This section requires your organization to "employ procedures and controls designed to ensure the authenticity, integrity, and when appropriate, the confidentiality of electronic records, and to ensure that the signer cannot readily repudiate the signed record as not genuine."

Basically, the requirement calls for:

- Proof that the change control record is authentic (date and time stamp)
- The record has not been tampered with (document encryption)
- Confidentiality has not been compromised
- The ability to discern altered records
- Authenticity of the electronic signature (use of digital signatures)

Section 11.30

This section requires companies using an open system to put procedures and controls in place, including:

- Document encryption
- Digital signatures
- Log-in controls, including password protection and password management

PROVING PART 11 COMPLIANCE

To prove that your electronic system complies with part 11 requirements you need to do the following:

- Develop process maps for the system: Initiation, evaluation, review, etc.
- Develop regulatory requirements as defined in part 11
- Develop system performance requirements: operating environment, scalability, etc.

- Develop functional requirements

These four bullets should lead you to develop your systems requirements matrix. The next step is to look for a system that can be configured to meet your requirements. Word of caution: there is no "one shoe fits all" when it comes to change control systems. Choose a configurable system that can meet your requirements. Design the system first; then look for the electronic solution, or infrastructure, to support it. Don't buy the software or system before you know what your requirements are. Lots of companies out there claim their systems are part 11-compliant. Put them to the test.

COMPUTER SYSTEM VALIDATION

The only way to prove that user requirements and part 11 requirements have been met is through validation. Whether the system is OTS (Off-the-Shelf), designed in-house, or custom-designed, it must be validated to prove that it is suitable for its intended use and that it meets part 11 requirements.

CHAPTER TEN:------------------------------System performance Indicators

It is essential to know how the change control system is performing. All improvements should be driven by facts and data. To accomplish this, the system performance matrix needs to be defined by the Quality Unit. Three matrices used to measure the performance of the change control system are:

- Cycle time
- Throughput
- Capacity loading

A. CYCLE TIME

Each requested change goes through a life cycle. The path taken depends on the decision made during the review process. Ideally, a requested change should go through the entire change control life cycle that your system defines. For example:

- Initiation phase
- Approval phase
- Implementation phase
- Effectiveness check phase
- Closure phase

  Essentially, a change should have and end in a linear manner. But as we know, life is not linear. Usually three problematic phases: the approval phase and effectiveness check. The two phases include cycles within the change control life cycle that, if unchecked, can be bottlenecks for approved changes, leading to delays and capacity management nightmares.

THE INITIATION AND REVIEW PHASE

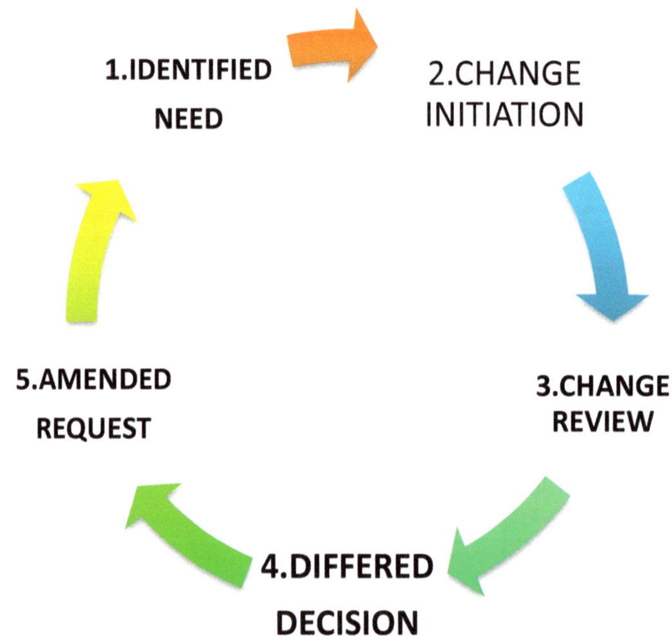

A Change review

The output of the review of a change request can be:

- Approved as-is
- Rejected
- Approved with amendments
- Deferred

A change request is deferred if the change control committee feels that more information is required for evaluation, or stakeholders need to be consulted before it is brought back to the committee. A deferred decision increases cycle time.

IMPLEMENTATION AND EFFECTIVESS CHECK

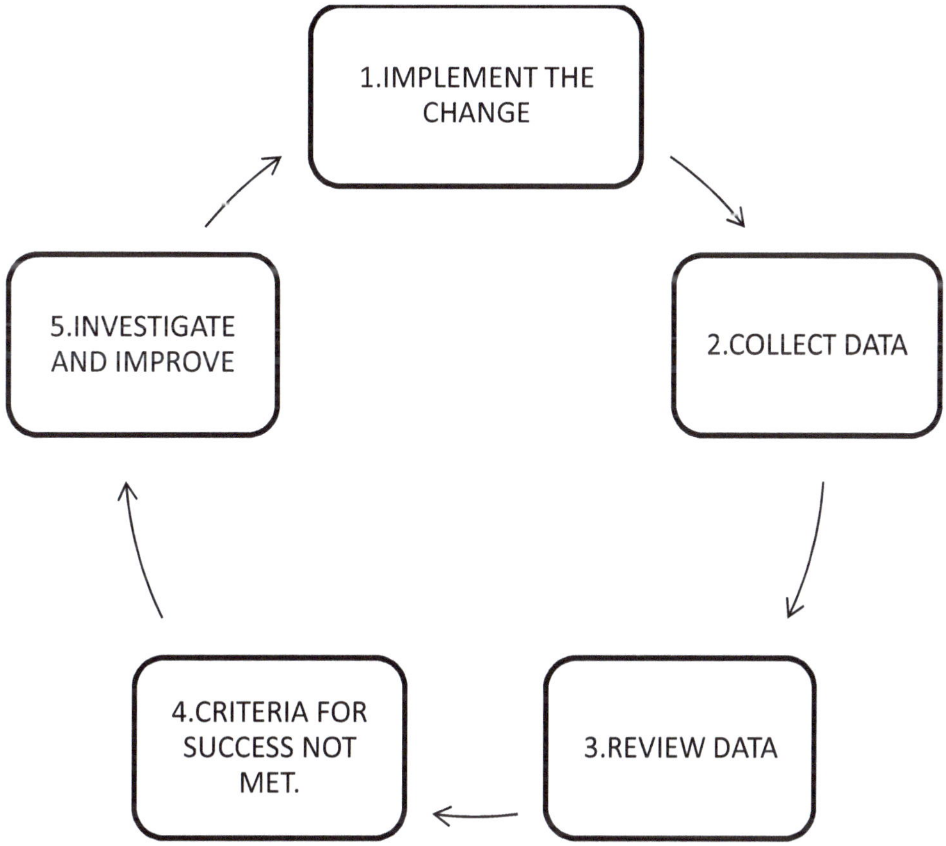

Causes of extensions in cycle time

These two circular loops often force change owners into requesting extensions for the change life cycle, hence reducing the process throughput and causing capacity management issues at implementation. The idea is to have a change control process with a predictable process cycle for minor, moderate, and major changes. Needless to say, the three categories of change requests should be defined.

B. THROUGHPUT

The throughput of a change control process refers to the number of changes that enter and leave the change control process within a chosen time frame. Examples here would include:

- Change requests processed per month
- Changes implemented per week
- Total Changes processed per year

C. CAPACITY LOADING

Capacity loading refers to the ability to process changes in the infrastructure. Capacity loading refers to the number of changes at every phase of the change control life cycle. The idea is to establish bottlenecks in the system and allow for efficient resource allocation. Capacity loading is a tool for capacity management for your change control process and the Quality system. It establishes the number of:

- change requests at Initiation
- changes in the review phase
- changes at implementation
- changes with extensions

The changes at each phase are further divided into:

1. Product
2. Process
3. Document
4. Infrastructure/IT
5. Equipment and facilities

The idea is to identify where most of the changes are coming from, to improve the Quality system. Repeated change requests from a manufacturing process would indicate a poorly designed manufacturing process that requires improvement. Continuous change requests about a marketed product would indicate weaknesses in product design, hence an indicator for product design improvements.

APPENDIX A---------------------------Definitions

The following terms, phrases and acronyms are used in change control.

Acronym	Definition
FMEA	Failure Mode and Effect Analysis: A method of assessing the effects of a risk being taken by implementing a particular change initiative
DFMEA	Design Failure Mode and Effect Analysis: A method used to assess the implications of a particular change in design specifications.
PFMEA	Process Failure Mode and Effect Analysis: A method used to assess potential risks with a change in process
FTA	Fault Tree Analysis: A methodology used to analyze implications of particular actions from the root
QSR	1. Quality System Regulations: Requirements in 21CFR 820.00 for a quality management system. The QSR requirements complement ISO9001 requirements. 2. Quality System Record
CGMP	Current Good Manufacturing Practices: FDA requirements for manufacturing medical devices, drugs, and biologics
CGTP	Current Good Tissue Practices: FDA requirements for tissue-processing organizations
SQEPP	Safety, Quality, Efficacy, Properties, and Potency
NIMBY	Not In My Back Yard
CR	Change Request: A form used in the initiation of a change
CC	Change Control: The act of assessing, approving, and implementing changes based on risks to product performance, process capability, the organization, and the end user
CCB	Change Control Board: A cross-functional group that meets to review change requests
CCC	Change Control Coordinator: A functional role for the person who coordinates change control activities. In some organizations this role is performed by a change control manager.
ECO	Engineering Change Order: The final order to execute an approved change request
ECN	Engineering Change Notice; Engineering Change Requests (ECR): Final notice for change implementation
SCAR	Supplier Corrective Action Request
RPN	Risk Priority Number: Ranking for risks involved in order of importance and agency. RPN= (O) X (S) X (D)
RI	Risk Index: Risk ranking
SOP	Standard Operating Procedure: Second-tier cross-functional documents
HAACP	Hazard Analysis And Critical Points

Term	Definition
Risk	Evaluation of the risks associated with the change request
Fit	External specifications and dimensions of a product
Form	Product methods for use, its appearance, and texture
Function	The product's performance and reliability
Performance	The product's ability to meet output specifications
Risk	Adverse effects
Deviation	Operating outside a set method or set specifications
Variance	A one-time deviation
Process	A set of steps that add value to an input to produce an output
Phase	High-level approach to activities that produce an output
Validation	Proof through objective evidence that the process or product consistently produces same results under pre-set conditions
Verification	Objective evidence that requirements were met
Compliance	Operating within set requirements
Change control	A systematic process of controlling risks associated changes related to products, processes, and quality management systems
Detection	Detective methods for stopping a failure mode
Evaluation	Objective review of pros and cons of a change request
Occurrence rating (O)	The likelihood of a failure taking place: On a scale of 1 to 3, 1 would mean less likely and 3 would mean very likely.
Detection rating (D)	Confidence in your detective methods: On a scale of 1 to 3, a rating of 1 would mean you are confident that current controls will catch the failure. A rating of 3 would mean you are not confident current controls will catch the failure.
Policy	High-level document that outlines the intent
Record	Proof that an activity was performed
Severity rating (S)	Consequence rating: On a scale of 1 to 3, a rating of 3 would mean disastrous, and 1 would be negligible.
Function	Reliability, quality, or appearance
Assessment	Evaluation process
Stakeholder	Department or functional role with vested interest in the proposed change
Purity	Absence of foreign material

Potency	Therapeutic properties of a drug as indicated by specified tests
System	A series of processes that act together as a unit to produce a desired and predetermined outcome
Safety	Freedom from harmful effects
Review	Activity performed to determine the suitability, adequacy, and effectiveness to achieve set objectives
Compliance	Meeting set or predetermined requirements
Specifications	Set requirements to which the product, process, or service must conform
Established	Defined, documented, and implemented
Component	Raw material, piece, or a part to be included as part of the finished product

Phrase	Definition
Change control system	Formal sequence of processes and steps through which changes related to product, process, and the quality system are handled to manage risks
Risk analysis	A formal methodology for analyzing the risks associated with a proposed change
Initiation phase	The first phase in the change control system
Evaluation phase	The approval phase of a change control system
Implementation phase	The third phase of a change control system
Closure phase	Final phase in the change control system. A change is considered closed if the effectiveness criteria are met and the quality unit has signed off.
Effectiveness check	The measure of success for a change request
Variance request	A one-time deviation request
Impact analysis	Evaluation of the financial impact of the proposed change on the organization
Capacity management	A method used to determine the number of different

	types of changes at a given phase in the system
Change request	A form completed by the initiator to start the change control process
Hazard assessment	Review of all possible hazards that may be associated with the proposed change
Responsibility matrix	A table that shows who does what
Stakeholder analysis	A breakdown of who may be affected and why they may be affected
Fault tree analysis	Graphical representation of a combination of factors or causes (roots) that may lead to a particular failure (effect)
Failure mode and effect analysis	A method of assessing the effects of a risk being taken by implementing a particular change initiative
Failure effect	The implications of a particular failure for the end user, the process, the organization, or regulatory standing
Failure mode	Manifestation of the failure (what it shows up as)
Design change control	Established and maintained procedures for identification, documentation, validation, or verification, review, and approval of design changes
Design review	Documented systematic examination of a design to evaluate the design to meet requirements
Design specifications	Product requirements established during design input. They are divided into three types: performance specifications, functional specifications, and non-functional specifications.
Change board	A cross-functional group that meets to review moderate and major change requests. It may include manufacturing, product development, engineering, quality, legal, and/or regulatory affairs.
Change control coordinator	A member of the quality organization who controls all change control activities. Some organizations have a change control manager.
Notification of changes	Supplier change notification notices
Process changes	Changes associated with specifications, methods, or procedures
Production changes	Deviations or variances
Document change records	Documented evidence of change description, identification of the affected documents, and approvals
Design changes	Changes associated with an approved design
Risk assessment	Evaluation of the risks associated with the change request. Usually evaluated in monetary values.

APPENDIX B-----------BIBLIOGRAPHY

FDA

Compliance Program Guidance Manual. *Transmittal 95-30.* May 4, 1995.

Deciding When to Submit 510(k) for a Change to an Existing Device. Draft 2. August 1, 1995.

Guidelines on General Principles of Process Validation. CDRH and Centers for Drug and Biologics, May 1987.

Inspection of Medical Device Manufacturers. *Compliance Program 7382.80, FDA Compliance Program Guidance Manual*. May 4, 1995.

Medical Device Quality System Manual: *A Small Entity Compliance Guide*. HHS Pub. No. 84-4191. April 1994.

Parts 210 and 211 cGMP in Manufacturing, Processing, Packing, or Holding of Drugs and Finished Pharmaceuticals.

21CFR Part 1271. Human Cells, Tissues, and Cellular Therapy and Tissue-based Products.

21*CFR Parts 600 and 606.* Blood Products and Biologics.

21CFR Part 820.00. Quality Systems Requirements.

21CFR 820.30. Design Control Guidance for Medical Device Manufacturers.

ISO

Quality Management Systems: Requirements. ISO 9001:2000.

Quality Systems: Medical Devices Supplementary Requirements to ISO 9001: 13485, 1995.

OTHER

Procedure for Performing a Failure Mode, Effects and Criticality Analysis. MIL-STD-1629A. November 1980.

Quality is everybody's job! ™

OTHER BOOKS BY THE SAME AUTHOR

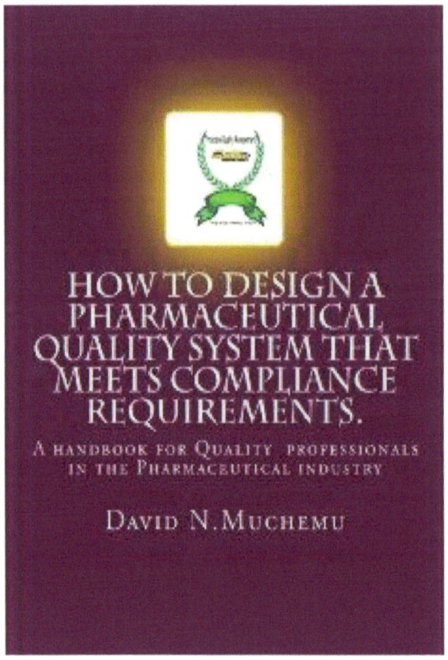

And many more! Available where books are sold and at http://www.cgmpuniversity.com

NOTES

QUALITY IS EVERYBODY'S JOB! ™

www.ingramcontent.com/pod-product-compliance
Lightning Source LLC
LaVergne TN
LVHW070937210926
848482LV00003B/260

9780615746395